The Bible:
MYTH
OR
DIVINE
TRUTH?

WARREN
HENDERSON

The Bible: Myth or Divine Truth? 4th edition
By Warren Henderson
Copyright © 2019
All rights reserved.

Published by Warren A. Henderson
3769 Indiana Road
Pomona, KS 66076

Cover Design by Benjamin Bredeweg

Perfect Bound ISBN 978-1-939770-50-9

eBook ISBN 978-1-939770-51-6

ORDERING INFORMATION:
Quantities of *The Bible: Myth or Divine Truth?* may be purchased through various online retailers or you may contact the author for pricing at:
https://www.warrenahendersonpublishing.com

Table of Contents

Acknowledgements

The author greatly appreciates all those who contributed to the publishing of *The Bible: Myth or Divine Truth.* Special thanks to Dr. Steve Ratering, Robert Sullivan, and Dr. Michael Windheuser for editorial contributions.

Video Presentation

A video presentation of materials contained in this book can be viewed at:

https://www.youtube.com/watch?v=CIxAkqP-hdY

Truth Doesn't Argue with Itself

What is real? What is truth? What is being? These are perhaps bizarre questions to some, but at the heart of these inquiries is the basis for how each one of us will interpret life. What an individual believes to be ultimate truth will define his or her ideology for relating to all of life's issues.

Human reasoning is vitally important to finding answers to life's most crucial questions. Thankfully, absolute truth cannot contradict itself. That which is not found consistently accurate cannot be true. It is understood that scientific perception of absolute truth is imperfect, for science is a process of refining its understanding of ultimate reality. On this fact, Albert Einstein once remarked, "One thing I have learned in a long life – that all our science, measured against reality, is primitive and childlike."[1] Science is certainly a viable path to what is observably true or conditionally true, but imperfections in the measurement, interpretation, and evaluation of data ensures that a scientific truth is only truth within certain ranges – a statistically valid interpretation of ultimate truth.

With this said, science is sufficient to validate, with a *high degree* of accuracy, what true reality is not. For example, if I say all ducks are black, but later a white duck is found, the antecedent is obviously wrong. If only black ducks were observed, it doesn't mean that a duck of a different color does not actually exist, we simply do not know if it exists; we have adequate confidence as to what is not true, but we cannot confirm what is true.

Consequently, if man wants to know the truth, he is going to have to look beyond science and world religion for answers. Empirically speaking, evidence indicates that the Bible is uniformly true and consistently accurate with scientific findings.

1

Offering evidence that supports this assertion is the focus of this book. Certainly there are many things in the Bible which cannot be proven true, but the questions to be explored are: (1) Is there any evidence which suggests that the Bible is not true? (2) Is there any evidence which validates the biblical narrative as true? Consider Archaeology for a moment; though Archaeology cannot directly prove that the Bible is divinely inspired, it does validate its historical reliability. If the Bible were divinely inspired, would not what science has learned from preserved history within the Earth, from the bountiful life on the Earth, and the splendor of celestial bodies beyond the Earth be in agreement? Christian author William MacDonald explains that truth cannot contradict itself: "No true finding of science will ever contradict the Bible, because the secrets of science were placed in the universe by the same One who wrote the Bible, God Himself."[2]

This book first presents evidences found in archeology, astronomy, cosmology, and various historical records which validate the Bible's accuracy. Secondly, internal evidence identifying the authenticity, uniformity, and prophetic content of the Bible is explained. Finally, a brief overview is provided of how religious "holy books," often stated to be of divine origin, hold up under the same scrutiny in which the Bible was examined by. There are no shades to divine truth; it is man who colors, flavors, changes, or dilutes truth in direct opposition to God's authority. God's Word has and will stand the test of time – it is immutable and absolute.

I believe that the evidence indicates that the Bible is consistently accurate with scientific discovery and with itself. There is no book on the planet with its authenticity, uniform composition, and prophetic content. If these assertions are true, it stands to reason that the Bible is an explicit expression of absolute truth to mankind. Reason then dictates that man should strive to learn of its message and heed its warning. If you have questions relating to what the Bible is and what God's central message to mankind is, please read on and discover the truth for yourself. God has preserved His personal message to you for thousands of years.

Archeology Agrees with the Bible

Prior to the 19th century, there was little archeological evidence which could validate the historical narratives within the Bible. Because there were civilizations, cities, sites, peoples, events, battles, kings, etc. mentioned in the Bible that could not be verified by external sources, a movement in the early 19th century was initiated to discredit the integrity of the Bible. Skeptics asserted that the Bible was merely a collection of folklore and myths.

Around this time, however, scientists began to dig beneath the surface of the Earth, and what they found silenced the critics. Hundreds of never before discovered cities, civilizations, and sites attested to the accurate narratives of the Old Testament.

If the Bible is God's revelation to mankind, would we not expect archeological findings to be consistent with the history recorded in Scripture? Yes; history is truth with a time reference. Although archaeology cannot directly prove the Bible's inspiration it does prove its historical reliability. On this point, Yale archeologist Millar Burrows writes:

> The more we find that items in the picture of the past presented by the Bible, even though not directly attested, are compatible with what we know from archaeology, the stronger is our impression of general authenticity. Mere legend or fiction would inevitably betray itself by anachronisms and incongruities.[1]

Fortunately, most of the claims of biblical legend and fiction have been silenced by numerous archeological findings. Many ancient civilizations, cities, and individuals recorded in the Bible have been verified by archeological evidence. The following are some of the more significant findings.

People, Cities, and Civilizations

Nebo-Sarsekim

In 2007, a Babylonian financial account was found among the British Museum's collection of 130,000 Assyrian cuneiform tablets, which documents the payment of 0.75 kg of gold to the temple in Babylon by Nebo-Sarsekim. The tablet is dated to the 10th year of the reign of Nebuchadnezzar II, 595 B.C., several years before the siege of Jerusalem. Jeremiah 39:3 identifies Nebo-Sarsekim as one of Nebuchadnezzar's generals present at the fall of Jerusalem in 587 B.C. Michael Jursa, an Assyriologist professor from Vienna, called the tablet the most important find in Biblical archaeology in one hundred years.[2]

Baruch

In 1975, the ancient clay Seal of Baruch, the son of Neriah (Jer. 36:4) was discovered. This archaeological find verifies the historicity of the book of Jeremiah. Baruch was the scribe who wrote down the prophecies of Jeremiah in about 607 B.C.[3]

Pontius Pilate

Various experts have maintained that Pontius Pilate was not an actual historical figure, but rather a fictional embellishment to the biblical account of Christ's crucifixion. However, in 1961 a slab stone of about 15 by 30 inches was unearthed at Caesarea Maritima that has the name "Pontius Pilate" ("Pontivs Pilatvs") inscribed on it.[4] The Pontius Pilate signet ring was found at the Herodium in 1968. In 2018, newly developed technology revealed that Pilate's name was inscribed on the ring.[5] The Roman historian Tacitus states that Pilate was Prefect of Judaea from 26 to 36 A.D. These findings prove that Pilate existed, as stated in the Bible's account of Jesus' civil trial (John 19).

The Hittites

The Bible records that the Hittites were a notable people in the Middle East from the days of Abraham (Gen. 15:20) to the time of King Solomon (1 Kgs. 10:29). Their legacy spans over a millennium. Until 1906, skeptics scoffed at the Bible's claim of the existence of the Hittites. However, archaeologists in the area of Annatolia, Turkey discovered the capital of the Hittite empire. A massive library of some ten thousand tablets (the royal archives of the Hittite Empire) dating back past 1600 B.C. was found.[6]

King David
Tel Dan Stela is a broken inscribed stone (stele) dating back to ninth-century B.C. It was found in 1993 during excavations of Tel Dan in northern Israel. An unnamed king (perhaps Syria's Hazael) boasts of victories over the king of Israel and his alley the king from the "House of David" (*BYTDWD*). This inscription provides evidence outside the Bible that King David existed.[7]

The City of Nahor
The Mari Tablets were discovered in 1933 along the Euphrates River, and contain the diplomatic correspondences and governmental records involving King Zimri-Lim (a contemporary of Babylon's Hammurabi). To date, some 20,000 clay tablets have been found dating back to about 1800 B.C. These tablets have biblical significance, such as naming the city of Nahor, which is identified in Genesis 24:10 and resided in the same region. The tablets mention a nomadic people, the "Habiru" or "Apiru;" this likely refers to the Hebrew people (Gen. 14:13).[8]

The Moabites
The Moabite Stone, which dates back to 850 B.C., was discovered by a German missionary in 1868 at Dibon, Jordan. This four feet high by two feet wide basalt monument documents Moabite attacks on Israel, as recorded in 2 Kings 1 and 3. Mesha was the king of the Moabites who was forced to pay tribute to Israel. The Bible states that the Moabite tribute suddenly stopped, *"Mesha, king of Moab, rebelled against the king of Israel..."* (2 Kgs. 3:5). Mesha's account of his rebellion against Israel is found

on this large stone. Prior to this finding, the existence of the Moabite people had not been confirmed.[9]

King Uzziah

King Uzziah ruled over Judah from 790 to 739B.C. Although mostly a good king, God struck Uzziah with leprosy for his intrusion of the priesthood (2 Chron. 26). After his death he was interred *"in the field of burial which belonged to the kings, for they said, 'He is a leper'"* (2 Chron. 26:23). His burial plaque was found on the Mount of Olives in 1931. It reads: "Here, the bones of Uzziah, King of Judah, were brought. Do not open."[10] Official seals for Kings Uzziah, Jeroboam, Hezekiah, and Hoshea have also been recovered at various excavation sites in Israel.[11]

Sodom and Gomorrah

Skeptics have long rejected the biblical account of God destroying the wicked cities of Sodom and Gomorrah by fire and brimstone. However, the 1974 discovery of the Ebla Tablets in northern Syria (Tell Mardikh) by Dr. Paolo Matthiae indicates that these cities did exist. To date, about 16,600 clay tablets have been unearthed from the Ebla Kingdom (2500-2250 B.C.). The tablets are written in Sumerian or in a cuneiform script of an ancient Semitic dialect. Dr. Giovanni Petinato found that some tablets referred to the city of Haran (Gen. 11:31) and to the five cities of the Jordan plain that were overtaken by four Mesopotamian kings in the time of Abraham (Gen. 14). One tablet actually lists these cities in the exact same order as Genesis 14:2 (Sodom, Gomorrah, Admah, Zeboiim and Zoar).[12] Furthermore, excavations at Jordan's Tall el-Hammam (likely the city of Sodom) have revealed that the city was suddenly destroyed by an intense incineration, which even transformed clay pottery into glass. Zircon crystals in the pottery, upon analysis, were shown to have formed within one second by superheating.[13]

Jericho

Original archaeological digs by Kathleen Kenyon in the 1950's found that Jericho's walls (except for a portion of the north wall) had not decayed over time, but rather suffered a massive collapse without rebuilding. After evaluating Kenyon's findings and doing further archaeological research Dr. Bryant Wood concluded, "… the city's walls could have come tumbling down at just the right time to match the biblical account [Josh. 6]. … It looks to me as though the biblical stories are correct."[14]

King Hezekiah and the Prophet Isaiah

The seal of Isaiah was found at the Ophel just south of the Temple Mount in 2009 (publicly revealed in 2018). The prophet likely used this seal to endorse legal documents (1 Kgs. 17-19). The official seal of King Hezekiah was also found at this site in 2015.[15] Moreover, 2 Kings 18 informs us of Hezekiah's religious reforms, including his removal of pagan shrines from Judah. Evidence found in 2016 at a Lachish gate shrine confirms that this purging occurred during Hezekiah's reign (716-687 B.C.).[16]

Events

Overthrow of Babylon

The Cyrus Cylinder was discovered in 1879 A.D. and records Persian King Cyrus' overthrow of Babylon and his subsequent release of Jewish captives. The Bible not only records these events in detail, but actually prophesied the name of Cyrus two centuries before his birth and how God would use him to defeat the Babylonian empire, release the Jewish captives and then assist them in rebuilding the temple in Jerusalem (Isa. 44:29-45:1).

Babylonian Siege of Jerusalem

The Lachish Letters discovered between 1932 and 1938 A.D. 24 miles north of Beersheba described the attack/siege of King Nebuchadnezzar on Jerusalem in 586 B.C.

Exodus and Canaan Conquest

The Amarna Tablets, discovered in 1887 at Amarna (a site along the Nile between Memphis and Thebes), record the military correspondences between Amenhotep IV (Pharaoh of Egypt) and his vassal state rulers in Canaan. The Canaanites were paying Pharaoh tribute in return for his protection. These tablets document the pleas for assistance from a Canaanite king named Abdi-Heba to Pharaoh against the invading "Habiru" (Hebrews). These roughly 300 tablets, dating back to 1300 B.C. provide credible evidence of the biblical account of the Hebrew exodus from Egypt and their later conquest of Canaan forty years later.[17]

The Rise of the Babylonian Empire

The Babylonian Chronicles were found in 1956; these four tablets date back to 600 B.C. These tablets record the historical events associated with the rise of the Babylonians: the conquering of the Assyrians, Egyptians and then Judah. All these events are well documented in the Bible (2 Kings, 2 Chronicles, Jeremiah, Ezekiel, and Daniel). The Babylonian Chronicles confirm military events such as the battle at Carchemish in 605 B.C. as recorded in Jeremiah 46:2 and 2 Kings 24:7-17.

Hezekiah's Water Tunnel

Edward Robinson discovered Hezekiah's tunnel in 1838. The tunnel runs for about 1750 feet (at a 0.6 percent grade) through solid rock at approximately 100 feet below the surface. According to the Bible, Hezekiah made a pool and a tunnel to bring water from the spring of Gihon into Jerusalem just prior to the Assyrian siege led by Sennacherib (2 Kgs. 20:20; 2 Chron. 32:30).

Verified Biblical Wars

Robert Sullivan of Christian Evidences Ministries compiled the following list of biblical wars which have been verified by extra-biblical evidence (which is noted after each account):

Campaign into Israel by Pharaoh Shishak (1 Kgs. 14:25-26), recorded on the walls of the Temple of Amun in Thebes, Egypt.
Revolt of Moab against Israel (2 Kgs. 1:1, 3:4-27), recorded on the Mesha Inscription.

Fall of Samaria (2 Kgs. 17:3-6, 24, 18:9-11) to Sargon II, king of Assyria, as recorded on his palace walls.

Defeat of Ashdod by Sargon II (Isa. 20:1), as recorded on his palace walls.

Campaign of the Assyrian king Sennacherib against Judah (2 Kgs. 18:13-16), as recorded on the Taylor Prism.

Siege of Lachish by Sennacherib (2 Kgs. 18:14, 17), as recorded on the Lachish reliefs.

Assassination of Sennacherib by his own sons (2 Kgs. 19:37), as recorded in the annals of his son Esarhaddon.

Fall of Nineveh as predicted by the prophets Nahum and Zephaniah (Nah. 2; Zeph. 2:13-15), recorded on the Tablet of Nabopolasar.

Fall of Jerusalem to Nebuchadnezzar, king of Babylon (2 Kgs. 24:10-14), as recorded in the Babylonian Chronicles.

Captivity of Jehoiachin, king of Judah, in Babylon (2 Kgs. 24:15-16), as recorded on the Babylonian Ration Records.

Fall of Babylon to the Medes and Persians (Dan. 5:30-31), as recorded on the Cyrus Cylinder.

Freeing of captives in Babylon by Cyrus the Great (Ezra 1:1-4, 6:3-4), as recorded on the Cyrus Cylinder.

Forcing Jews to leave Rome during the reign of Claudius (A.D. 41-54) (Acts 18:2), as recorded by Suetonius.[18]

Verified Biblical Locations and Structures

Besides civilizations, cities, individuals, and wars, archeological evidence has located hundreds of biblical sites. Consider the following abbreviated list: the Palace at Jericho. the East Gate of Shechem, the Temple of Baal/El-Berith in Shechem, the Pool of Gibeon, the Pool of Heshbon, the Palace at Samaria, the Pool of Samaria, the Palace in Babylon, the Palace in Susa, the Royal Gate at Susa, Jacob's well, the Pool of Bethesda, the Pool of Siloam, the Theater at Ephesus, and Herod's palace at Caesarea.[19]

Summary

Archaeology cannot directly prove the Bible's inspiration, but it has verified its historical reliability by identifying numerous ancient sites, civilizations, and individuals which the Bible identifies. Unearthed treasures, such as the Cyrus Cylinder, the

Mari Tablets, the Nuzi Tablets, the Moabite Stone, the Ebla Tablets, the Amarna Tablets, the Lachish Letters, Sennacherib's Prism, the Ras Shamra Tablets, and the Babylonian Chronicles have all confirmed the biblical record.

Just how strongly does archeological evidence affirm the accuracy of the Bible? Consider what the experts have to say:

> On the whole, however, archaeological work has unquestionably strengthened confidence in the reliability of the scriptural record. More than one archaeologist has found his respect for the Bible increased by the experience of excavation in Palestine.[20]
>
> — Millar Burrow, Yale Archaeologist

> No fact of archaeology so far discovered contradicts the Biblical records.[21]
>
> — Donald Wiseman, Professor Emeritus of Assyriology at London University

> No archaeological discovery has ever controverted a biblical reference.[22]
>
> — Nelson Glueck, renowned Jewish archaeologist

The conclusion is obvious. The Bible is not a collection of mythical stories as some claim, but rather, it is an accurate rendition of ancient history. As truth will not contradict itself, all that is hidden in depths of the Earth will be a witness to the integrity of Scripture. If there is an apparent contradiction between the Bible and archeological findings, it is the interpretation of those findings which are suspect and not the accuracy of Scripture. Dr. Frank Harber explains that this is not a presumptuous conclusion, but a logical one:

> Over 25,000 sites have been discovered which have a connection with the Old Testament period. Not only have these discoveries provided external confirmation to hundreds of scriptural assertions, but not one single archaeological discovery has ever contradicted a biblical assertion.[23]

Science Agrees with the Bible

Those who consider nature to be all there is are referred to as *naturalists*, while those who believe that something else exists besides nature are called *supernaturalists*. Naturalists believe only nature itself is real; it is everything and the cause of everything. Though supernaturalists have varying beliefs, they do agree that there was a conscious initial cause that resulted in the natural order we witness today.

Even though the naturalist and supernaturalist have incompatible worldviews for interpreting life, both have a common fallibility when interpreting science to explain the existence of life. Both the supernaturalist and the naturalist impose their personal beliefs into the evaluation process – pure objectivity in science is unattainable. Sir Arthur Eddington recognized this tendency in scientists: "In science we sometimes have convictions which we cherish but cannot justify; we are influenced by some innate sense of the fitness of things."[1] Whether we like to admit it or not, scientific conclusions are clouded by personal bias as to how we want to interpret the meaning of life.

Man has developed systems of logic, namely *induction*, *deduction*, *abduction*, and various integral forms of these to explore absolute truth to the degree feasible. Reason, statistical analysis, observation, demonstration, mathematical modeling, etc. all have their part in man's quest for determining the truth about himself and his surroundings. The limitation with each of these synthetic logic systems is that each originates with man and is thus governed by imperfect reasoning (induction, abduction, deduction, falsification, etc. are imperfect reasoning systems). As man cannot naturally know absolute truth, it stands to reason that man's devices to pursue truth will be imperfect also. The conclusion of a deductive argument is only true if the premises are perfect; but how does one validate perfection? If a divine being who truly transcends natural order does exist, any imperfect reasoning system constrained by natural law would then be inconclusive in validating His existence (my book *In Search of God* provides a lengthy proof of this statement).

In summary, science narrows in on absolute truth to the degree that imperfect observation, imperfect verification, and inserted human bias can allow. Perfect scientific conclusions based on empirical evidence are not possible because science cannot reason results perfectly. With this limitation stated, let us review what we perceive to be true from scientific evidence in correlation with what the Bible states to be natural reality.

Science Confirms Biblical Truth

If the Bible is divine in origin then we would expect science to both detect and confirm with a high degree of accuracy biblical statements concerning nature. The following is an abbreviated list of biblical points concerning nature. Though scientists do not agree on the initial cause of nature or the timing of its realization, many would agree with the following biblical insights, whether he or she were a creationist or a naturalist. For example, those holding to the Big Bang theory could agree that….

1. All that is visible came from what was invisible (Heb. 11:3).

2. Before creation, time did not exist (1 Cor. 2:7; 2 Tim. 1:9).

3. Universe continues to expand (Job. 9:8; Ps. 104:2; Isa. 44:24).

4. Universe is winding down and will eventually "wear out" (Ps. 102:25-26).

5. The universe has innumerable number of stars (Gen. 22:17; Jer. 33:22). Note: only about 3000 can be seen with naked eye.

6. Every star is unique, differing in magnitude (1 Cor. 15:41).

7. "In the beginning (time), God created the heavens (space) and the Earth (matter)" (Gen. 1:1). The relationship of time, space, and matter is acknowledged in the first verse of the Bible.

8. That there was an order of biological development from lifeless matter to sophisticated life-forms (Gen. 1). Though naturalists will likely reject the idea of distinct supernatural acts of creation; most would agree with the Genesis 1 order of biological development. The Bible states that God created something out of nothing (Gen. 1:1), then created unconscious life – vegetation (Gen. 1:11), then conscious life – sea life, birds, land creatures (Gen. 1:20-21, 25) and lastly human life (Gen. 1:26-27).

The Bible supplies a number of accurate scientific statements about the physical properties of nature which are witnessed in the universe. For example, the Bible upholds the fact that matter can neither be created nor destroyed (the First Law of Thermodynamics) and the universal law of system decay (the Second Law of Thermodynamics). Genesis 2:1 declares that God "finished" all that was to be created, therefore, until God chooses to destroy creation, the natural order He put in place shall be maintained. Neither matter nor energy can be either created or destroyed after creation. Likewise the entropy (a measurement of system disorder) of all creation cannot decrease, it can only increase or remain the same. As the Psalmist notes: *"Of old You laid the foundation of the Earth, and the heavens are the work of Your hands, they will perish, but You will endure; yes, they will all grow old like a garment"* (Ps. 102:25-26; see also Heb. 1:10-11).

Besides the six-day creation account (Gen. 1), the Bible reveals the Creator's connection with His creation. God is a God of order, not confusion (1 Cor. 14:33); He creates with exact purpose. God created the world by His power and wisdom (Jer. 10:12). Man was created in God's image (to represent God on Earth) and His likeness (to morally behave like God; Gen. 1:26). Both aspects were lost when man disobeyed God.

The Bible teaches that God created man, not that man evolved (Gen. 1:27) and that He created male and female from the beginning (Matt. 19:4). Therefore, the theory of "theistic" or "God-guided" evolution would not be upheld by the Bible. How does scientific evidence align with the Bible's account of natural law? The following evidence is arranged by scientific discipline.

13

Astronomy

The Sun has a Circular Path

"Its [speaking of the sun] rising is from one end of heaven, and its circuit to the other end; and there is nothing hidden from its heat" (Ps. 19:1-6). Skeptics once asserted that these verses were incorrect; the sun doesn't travel on a circular path – it's just the planets that do. However, we now know that the sun moves in a "circuit" or circular course through space, at speeds approaching 600,000 miles per hour, near one of the spiral arms of the Milky Way Galaxy. Furthermore, the Milky Way is hurtling through space at an estimated speed of 2,000,000 miles per hour.[2]

The Earth is Circular

"It is He who sits above the circle of the Earth..." (Isa. 40:22). The Hebrew word *chuwg* implies a continuous circle or circuit about the Earth, some have suggest this meaning to convey the thought of a sphere, in any case, a circular shape is implied by the text. It is noted that some skeptics have pointed to the *"four corners of the Earth"* statement in Revelation 7:1 as evidence that the Bible states the Earth has corners. The Greek word translated "corners" is *gonia* and is understood to mean a literal "corner" or a regional "quarters." Most revelation in the prophetic books of the Bible was given audibly, but we understand that the means of conveying information in Revelation would be mainly through viewed symbols and figurative scenes (Rev. 1:1). Thus, the reference simply implies a worldwide consequence – the four quarters of the Earth. Four is the number symbolically used for earthly order throughout Scripture, thus the numeric meaning is held consistent in the context of the passage.

The Earth Hangs on Nothing

"He stretches out the north over empty space; He hangs the Earth on nothing" (Job 26:7). Throughout the centuries philosophers, theologians and scientists have theorized as to how the Earth was supported. The ancient Greeks asserted that either Atlas or Hercules shouldered the planet. The Hindu *Vedas* teaches

that the Earth is flat, triangular in shape, and held up on the back of four elephants (perhaps turtles). Yet, the concept of a free-floating planet hanging in nothingness was a foreign concept to ancient man, who was subject to gravity. The Bible rightly states that the Earth hangs in empty space.

Pleiades and Orion

Job may be the oldest book in the Bible (about 3500 years old). In correcting Job, God poses this question, *"Can you bind the cluster of the Pleiades, or loose the belt of Orion?"* (Job 38:31). We now understand what God was sharing with Job. Within the Taurus constellation is a tight grouping of stars in gravitation lock; they are called "Pleiades." Although many stars are in this cluster (about 440 light years away), only seven are discernable with the naked eye on a clear night; sometimes these are referred to as the "seven sisters." Just as the Bible states, these stars are bound together; they cannot pull apart from one another. However, the constellation Orion is composed of stars throughout our galaxy, and we know that the Milky Way is expanding. As the years roll by, Orion's belt is literally *letting out a notch*. The answer to God's question to Job was that only God can arrange and control the constellations in such a way that He binds some stars together and loosens others.

Cosmology
The Water Cycle

Solomon, Isaiah and Job all describe the evaporative cycle. Solomon wrote, *"All the rivers run into the sea, yet the sea is not full; to the place from which the rivers come, there they return again"* (Eccl. 1:7). Job declares, *"For He draws up drops of water, which distill as rain from the mist, which the clouds drop down and pour abundantly on man"* (Job 36:27-28, also 26:8). The hydrologic cycle was not discovered until the late seventeenth century when Pierre Perrault and Edme Marriotte proved that rain water was sufficient to sustain river flow and Edmond Halley further demonstrated that oceanic evaporation supplied the moisture for rain clouds which supplied river waters.[3]

The Weight of Air

It was not until after Italian physicist Evangelista Toricelli constructed the first barometer in 1643 that atmospheric pressure was measured. Toricelli noted that small changes in the mercury's height in his barometer directly correlated with changing weather conditions.[4] He had proven that the air above us exerts a pressure on us because air has mass. Yet, the Bible stated long ago that air has weight, *"For He looks to the ends of the Earth, and sees under the whole heavens, To establish a weight for the wind, and apportion the waters by measure"* (Job 28:24-25).

Oceanography

Paths in the Sea

Some 3000 years ago, David wrote concerning *"the paths in the sea"* (Ps. 8:8). This verse prompted 19th century naval officer Matthew Fontaine Maury to explore the oceans for prevailing sea currents. After circumnavigating the world from 1832 to 1836, he published the first textbook on modern oceanography, *The Physical Geography of the Sea and Its Meteorology*. Maury, indeed found "the paths in the sea," pioneered the science of oceanography, and revolutionized the trade routes for sailing ships.

Valleys in the Sea

Within approximately the same time frame that David wrote of "the paths in the sea," the book of Samuel recorded that valleys or channels existed in the floor of the sea (2 Sam. 22:16). Until modern times people actually believed that ocean floor was a sandy saucer shaped desert, and thus the ocean was deepest in the middle. In early 20th century oceanographers discovered that the sea had many deep valleys or trenches. The Marianas Trench in the Pacific Ocean is so deep that if Mt. Everest was dropped into it, the peak would still be a mile below the water's surface.[5]

Mountains in the Sea

Did you know the tallest mountain on the planet is not Mt. Everest? Although the Everest peak is the highest location on

16

Earth (rising some 29,035 feet above sea level) it is not the highest mountain on the planet; the honor belongs to Hawaii's Mauna Kea. Mauna Kea is observed to be only 13,796 feet above sea level, but it begins rising off the ocean floor nearly three and a half miles below the ocean's surface. Mauna Kea, when measured from the ocean floor, reaches a height of 31,796 feet. It is one of many mountains extending up from the ocean's floor and one of the few viewed above the ocean's surface.

Mauna Loa is also found in the Hawaiian Islands and rises some 13,677 feet above sea level. Because of its vast base, Mauna Loa is considered the most voluminous mountain on the Earth.[6] Much of the sea floor has yet to be explored, but we now know that mountains and even mountain ranges are plentiful (the Atlantic Ocean contains an undersea mountain range 10,000 miles long). Jonah declared long ago what modern science is just now understanding, *"The waters surrounded me, even to my soul; the deep closed around me; weeds were wrapped around my head. I went down to the moorings of the mountains"* (Jon. 2:5-6).

Springs in the Sea

Job had questioned God's wisdom given his present distress and suffering. God taught Job about divine sovereignty through a series of questions. One such question was, *"Have you entered the springs of the sea?"* (Job 38:16). These oceanic springs were not discovered until Matthew Fontaine Maury demonstrated their existence in the 18th Century. However, firsthand proof did not exist until 1977, when Lewis Thomas (*The Medusa and the Snail*) documented that scientists had found springs in the ocean floor off the coast of Ecuador, at a depth of 1.5 miles. It is estimated that 40 cubic miles of hot water pours into the Earth's oceans each year from these oceanic vents – "hot springs."[7]

Meteorology
Jet Streams

"The wind goes toward the south, and turns around to the north; the wind whirls about continually, and comes again on its

circuit" (Eccl. 1:6). Jet streams are narrow channels of fast flowing air, typically about six miles above the surface of the Earth, which form where air masses of differing temperature gradients collide. In the northern hemisphere the jet streams flow from west to east, but can widely vary in north to south orientation. Air speeds of 35 to 75 miles per hour are common. It was not until airmen were doing bombing runs over Germany and Japan that these high speed streams of air were better understood.

Paleontology

Does paleontology support the biblical record of creation or the theory of evolution? Vance Ferrell's *The Evolution Handbook* (2006) is an excellent compilation of information on evolution. He draws from the testimonies of hundreds of experts in various scientific fields and summarizes key problems with evolutionary theory as based on the fossil evidence found in various strata. He states, "These problems are serious enough that any one of them is enough to overthrow the evolutionary theory in regard to paleontology and stratigraphy." How does the fossil record oppose the theory of evolution?

1. Life suddenly appears in the bottom fossil-strata level, the Cambrian, with no precursors.
2. When these lowest life forms appear (they are small slow-moving, shallow-sea creatures), they are outstandingly abundant, numbered in the billions of specimens, and quite complex.
3. No transitional species are to be found at the bottom of the strata, the Cambrian.
4. Just below the Cambrian, in the Precambrian, there are no fossil specimens [other than some surface algae and a few debated protists and bacteria].
5. No transitional species are to be found below the lowest stratum, in the Precambrian.
6. No transitional species are to be found above the bottom stratum, from the Ordovician on up.
7. Higher taxa (forms of life) appear just as suddenly in the strata farther up. These higher types (i.e., beavers, giraffes, etc.) suddenly appear with no hint of transitional life forms leading up to them.

8. When they appear, vast numbers of these life forms are to be found.[8]

Most of us, from early grammar school upwards, were taught that evolution was the best scientific explanation of life. Was this explanation, however, based on true scientific findings or conjecture? Does the fossil record uphold evolution as a valid theory? What is the consensus of leading evolutionists concerning the fossil record?

The absence of fossil evidence for intermediary stages between major transitions in organic design, indeed our inability, even in our imagination, to construct functional intermediates in many cases, has been a persistent and nagging problem for gradualistic accounts of evolution.

> — Steven J. Gould, "Is a new and general theory of evolution emerging?", Paleobiology 6:119–130, 1980, p.127.

The fact is that subsequently no new phyla have appeared, and no new classes and orders. This fact, which has been long ignored, is perhaps the most powerful of all arguments against Darwin's generalization.

> — G. R. Taylor, in *The Great Evolution Mystery* (1983), p. 138.

Paleontology is now looking at what it actually finds, not what it is told that it is supposed to find. As is now well known, most fossil species appear instantaneously in the record, persist for some millions of years virtually unchanged, only to disappear abruptly.

> — Tom Kemp, "A Fresh Look at the Fossil Record," *New Scientist* (1985), p. 66.

It is a simple ineluctable truth that virtually all members of a biota [regional plant and animal life] remain basically stable, with minor fluctuations, throughout their durations [speaking of the fossil record].

> — Niles Eldredge, *The Pattern of Evolution* (W. H. Freeman, New York; 1998), p. 157.

> Wherever we look at the living biota ... discontinuities are over-whelmingly frequent.... The discontinuities are even more striking in the fossil record. New species usually appear in the fossil record suddenly, not connected with their ancestors by a series of intermediates.
>
> — E. Mayr, *What is Evolution*; 2001, p. 189.

Although many of the experts in the field of paleontology still hold to some form of evolutionary thinking, their keen evaluations of the fossil record indicate serious problems with the theory. The general consensus of paleontologists has not changed much since Darwin first published *The Origin of Species* – the testimony of fossils does not favor the theory of evolution. Yet, with this said there is nothing within the fossil record that is contrary to the Bible's creation or world-wide flood accounts.

Other naturalists beyond paleontology have also questioned Darwinian science. Biologist Lynn Margulis (former wife of Carl Sagan) concluded in 2006: "new mutations don't create new species; they create offspring that are impaired." "Neo-Darwinists say that new species emerge when mutations occur and modify an organism. I was taught over and over again that the accumulation of random mutations led to evolutionary change – led to new species. I believed it until I looked for evidence."[9]

In 2011, R. C. Tallis (prof. of medicine at Univ. of Manchester), a self-proclaimed "atheist humanist," critiqued the scientism invading the study of consciousness and the origin of the human mind: "Darwinism, therefore, leaves something unaccounted for: the emergence of people like you and me ... Isn't there a problem in explaining how the blind forces of physics brought about (cognitively) sighted humans who are able to see, and identify, and comment on, the 'blind' forces of physics...?"Tallis admits "the failure to explain *any* form of consciousness, never mind human consciousness, in evolutionary terms."[10]

The reader must determine for himself or herself what account of the existence of nature and the origin of humanity better aligns with the testimony of science. The Bible contains dozens of accurate descriptions pertaining to creation order and these are in perfect agreement with scientific evidence.

Is Jesus a Historical Figure?

It is a common ploy of some atheists not only to deny the authenticity and central message of the Bible (i.e. Jesus Christ, the Son of God, came to Earth to seek and save the lost), but to further assert that Jesus Christ of Nazareth never existed. For example, Bertrand Arthur Russell wrote in his book *Why I Am Not a Christian*: "Historically, it is quite doubtful whether Christ ever existed at all…" Skeptics, like Russell, claim that there is no external evidence that confirms the Bible's extensive account of the life of Jesus Christ. Is this a valid conclusion, or is there external evidence which verifies that a person named Jesus from Nazareth actually existed? Judge for yourself.

Writing shortly after the time of Christ's death, Jewish historian Flavius Josephus wrote: "Now, there was about this time Jesus, a wise man, if it be lawful to call him a man, for he was a doer of wonderful works – a teacher of such men as received the truth with pleasure."[1] Conveying the story of the stoning of James the half brother of Christ, Josephus wrote: "So he [Ananus the High Priest] assembled the Sanhedrin of judges, and brought before them the brother of Jesus, who was called Christ, whose name was James…."[2] The reference to James, a leader in the early Church and the half brother of Christ, agrees with Scriptures' reckoning (Matt. 13:55; Mark 6:3; Acts 15:13; Gal. 1:19). Josephus was referring to a genuine historical Jesus.

Some have asserted that Josephus' statements were "refined" by a Christian editor centuries later. While this is possible, it doesn't undermine that Josephus refers to a historical Jesus. Furthermore, in 1972, a Jewish scholar found a copy of the above passage in an Arabic translation of Josephus's Greek writings. The original translator was most likely a Moslem. Since neither medieval Jews nor medieval Moslems would have had any reason to authenticate the historic life of Jesus, the mention of Jesus in Josephus's writings appears authentic.[3]

Reporting on Emperor Nero's decision to blame the Christians for the fire destroying much of Rome in A.D. 64, the Roman historian Tacitus wrote:

> Nero fastened the guilt … on a class hated for their abominations, called Christians by the populace. Christus, from whom the name had its origin, suffered the extreme penalty during the reign of Tiberius at the hands of … Pontius Pilatus, and a most mischievous superstition, thus checked for the moment, again broke out not only in Judaea, the first source of the evil, but even in Rome ….[4]

A third important historical source of evidence concerning Jesus Christ and those who believed on Him is found in the letters of Pliny (the younger), a Roman governor of Bithynia (then a province of Asia Minor) to Emperor Trajan asking for advice on how to legally prosecute those accused of being Christians. In one letter (about 112 A.D.), Pliny relates to Trajan what he has learned about these Christians (who were numerous in Bithynia):

> They were in the habit of meeting on a certain fixed day before it was light, when they sang in alternate verses a hymn to Christ, as to a god, and bound themselves by a solemn oath, not to any wicked deeds, but never to commit any fraud, theft or adultery, never to falsify their word, nor deny a trust when they should be called upon to deliver it up; after which it was their custom to separate, and then reassemble to partake of food – but food of an ordinary and innocent kind.[5]

Lucian of Samosata was a second century Greek satirist. While his flippant remarks in one of his works were meant to ridicule early Christians, he does supply significant comments about their founder Jesus Christ. He wrote:

> The Christians … worship a man to this day – the distinguished personage who introduced their novel rites, and was crucified on that account. … [It] was impressed on them by their original

lawgiver that they are all brothers, from the moment that they are converted, and deny the gods of Greece, and worship the crucified sage, and live after his laws.[6]

Suetonius mentions Christ in his biography of the Roman Emperor Claudius and repeatedly observes, in his biographies of Caligula and Vespasian, that the Romans knew about Christ's prophecy of a King rising in the East who would rule the entire world.

The Talmud (a collection of Rabbinical writings from 70-500 A.D.) verifies Jesus' existence by attempting to discredit Him:

On the eve of the Passover Yeshu was hanged. For forty days before the execution took place, a herald ... cried, "He is going forth to be stoned because he has practiced sorcery and enticed Israel to apostasy."[7]

"Yeshu" (or Yeshua) is Jesus' name in Hebrew. According to the Talmud, Jewish authorities believed that Jesus was an apostate; this agrees with the biblical account (Matt. 26:65-66). He was "hanged," which is a term the Bible repeatedly applies to crucifixion (Luke 23:39; Gal. 3:13). The Jews normally stoned condemned criminals, but being under Roman rule, they had to yield to Roman execution methods (John 18:31-32). Yeshu was therefore crucified as foretold in Psalm 22:16-18. The Talmud also states that Yeshu died on the eve of Passover (John 19:14).

Besides Josephus, Tacitus, Pliny, Lucian, Suetonius, and the Talmud, other non-biblical first century writers have documented the unique life of the Lord Jesus Christ. Men such as Clement of Rome (a leader in the Church at Rome and likely a companion of Paul), Hermas (likely referred to by Paul in Romans 16:14), Ignatius (a personal disciple of one or more of the original apostles), Papias (a disciple of John the apostle), and Polycarp (a disciple of John the apostle). All these men had access to information pertaining to Jesus Christ from eyewitness accounts.

Besides these, there are ample second century writings still in existence which collaborate the biblical accounts of the life of

Christ. These accounts are from writers such as: Apollonarius, Aristides, Athenagoras, Celsus, Clement of Alexandria, Dionysius of Corinth, Hegesippus, Hippolytus, Irenaeus, Julius Africanus, Justin Martyr, Marcion, Melito, Montanus, Polycrates, Tatian, Tertulian, and Theophilus.

Summary

Both the historical and the biblical evidence is overwhelming to substantiate the unique teachings, miracles, and character of Jesus Christ. He really died after being nailed to a Roman cross and arose from the grave three days later just as He said He would.

Clement of Rome wrote just a few years after Christ's resurrection, "God has made the Lord Jesus Christ the first-fruits by raising Him from the dead."[8] His contemporary Ignatius recorded this statement: "For I know that after His resurrection also, He was still possessed of flesh. And I believe that He is so now."[9] The external evidence is quite conclusive; Jesus of Nazareth was a historical figure, just as the New Testament narrative declares!

Biblically speaking, Jesus Christ proved His credentials as Messiah and Savior, the incarnate Son of God by working many miracles – each one a sign to the Jews and a warning if rejected. Jesus Christ changed, controlled and multiplied the elements, healed the sick, rebuked evil spirits, raised the dead, and affected His own death and resurrection. His own resurrection would be the ultimate sign to the Jews and would ultimately determine whether or not they would believe on Him as Messiah (Matt. 12:38-40). If there was no resurrection, then Jesus Christ obviously could not be their Messiah.

Jesus Christ truly was who He proclaimed to be and His miracles, including His own resurrection, provide evidence to this truth. The external evidence supports the Bible's presentation of Jesus Christ in the Gospels. There is no logical middle ground on this matter; as Thomas Aquinas surmised centuries ago, "Christ was either liar, lunatic, or Lord!"[10] So which is it? If Jesus Christ is not Lord of all, He is not Lord at all.

The Authenticity of the Bible

There simply is no ancient book on the planet that has the same authenticity as the Bible. Scientific analysis has verified that its original text has been almost perfectly preserved down through the centuries. When one considers how minute the Jewish nation was in comparison to surrounding nations and in respect to past world empires, it simply is amazing that the Scripture entrusted to the Jews has survived and not volumes of religious writings from the larger pagan populace. The Jews were a scattered people from 605 B.C. until 1948 A.D., yet the Old Testament was incredibly preserved, a testimony to their diligence to keep it.

If the Bible is God's word to mankind, would He not maintain an accurate record of it through the corridors of time? Without Scripture we would not know Christ, the Savior, the divine solution for man's sin. This explains why the Bible, God's beacon of truth to mankind, has been so well preserved. God desires those in future generations to know Him and be restored to Him.

Some have said you cannot trust the integrity of the Bible. The Bible, however, is the most widely examined book on the planet; its textual integrity has been repeatedly validated. When copying worn out scrolls the Jews verified the integrity of each new scroll by counting each letter of the Hebrew alphabet and comparing it with the master's letter count; the original was destroyed only after an exact copy was verified. How can I know for sure that the Bible is God's message to humanity and that it is still trustworthy? Please carefully consider the evidence in this chapter and the two subsequent chapters.

When speaking of a book that is as much as 3,500 years old, it is understood that the original autographs would be unavailable, having perished over time. Today, the nearest fragment of Scripture to its autograph is dated within one generation. Scripture in its original form is God-inspired and perfect (2 Tim. 3:16). From a rational standpoint, this does not prove that the Bible is God's Word, for that would be circular logic that any religious book might self-proclaim – the book states it is revelation from God; therefore, it must be. The point here is to substantiate the internal evidence – the Bible declares that it is God's word to mankind.

It is noted that before the 15th century, Bibles were not printed but hand copied onto papyrus, parchment and later paper; these copies are called manuscripts. Over time, a few scribal errors were noticed (e.g. the copying of Hebrew numbers for example), but most of these can be identified and eliminated by comparing the various manuscripts with each other (thousands are available to accomplish this). As a result of such comparison, 99.5 percent of the New Testament has been determined to be authentic to the original autographs.[1]

Those who criticize the integrity of the Bible, such as those of the *Jesus Seminar*, toss about staggering figures claiming that there are over 200,000 variant readings in the existing New Testament manuscripts. What they don't tell you, however, is that if one single word is misspelled in 3,000 manuscripts, it is counted as 3,000 variants! Nearly all variants are simple spelling variations, a lost letter or two words transposed. It doesn't take long to add up to 200,000 with this type of reckoning.[2] In other words, more than 99 percent of the variants in the New Testament are not noticeable when the text is translated; of the remaining differences, none affects any vital aspect of the Christian faith.[3]

Concerning Old Testament authenticity, the finding of the *Dead Sea Scrolls* in 1947 has proven most significant. Some 900 separate documents and approximately 40,000 fragments dating between 300 B.C. and 70 A.D were found (these contain portions of every book in the Old Testament excluding Esther). Analysis of all available documents show that approximately 90 percent of the Old Testament text is without any variation. The variances in the remaining ten percent are insignificant (most variances relate to an obvious slip of the pen, inadequate knowledge of the Hebrew language by the translator, or dialect differences).[4] When proper textual criticism is conducted, a textual purity exceeding 95 percent results.[5] The remaining uncertainties mostly amount to a simple discrepancy in word order, which when accounted for infer an even higher accuracy. As the truth of Scripture is in the whole, the small percent remaining is best tested by the integrity of Scripture itself – the result of which indicates that no new doctrines are imposed by these minute manuscript variations.

The differences between the oldest known Old Testament Hebrew text (800 A.D.) prior to the *Dead Sea Scrolls* discovery and the actual *Dead Sea Scrolls* were found to be quite minor? For example, Isaiah chapter 53 is an astounding prophetic passage in the Old Testament that describes why Messiah must suffer, die, and be raised up again. Examining the 166 words in Isaiah 53, we find only 17 letters in question. Ten of these are simply a matter of spelling variations; four pertain to minor stylistic changes, and the remaining three Hebrew letters compose one word, "light," apparently added in verse 11 to the Masoretic text – which does not affect the meaning of the passage at all.[6]

Other prominent manuscripts include the *Geniza Fragments* (400 A.D.) and the *Ben Asher Manuscripts* (950 A.D.). As man continues to dig, the authenticity of the Bible becomes more apparent. For example, in 1979, archaeologist Gabriel Barkay found two small silver scrolls dating back to 600 B.C. in a Jerusalem tomb. The scrolls contained a benediction from the Book of Numbers which proved that the Old Testament had already been copied at a time when skeptics thought the text didn't yet exist.[7]

What is the oldest New Testament manuscript in existence today? The *John Rylands Fragment*, dating to 117-138 A.D. is a papyrus piece written on both sides; it includes John 18:31-33, 37-38. Dr. Carsten Thiede dates the *Magdalen Papyrus* containing verses from Matthew 26 to about 70 A.D.; others favor a second century date.[8]

The entire New Testament was in circulation by 165 A.D.; the Muratorian Canon of that era excluded, Hebrews, James, 1 and 2 Peter, and 3 John. Over 5,600 Greek partial or complete New Testament manuscripts exist today which attest to the reliable and accurate transmission of Scripture. When you consider how few important Greek texts have survived 2000 years of history, one must recognize that the Bible is God's Word and that He has preserved His Word. Norman Geisler notes how few copies of important Greek texts have survived two millennia:

- Nine or ten good ones for Caesar's Gallic Wars.
- Twenty manuscripts for Livy's History of Rome.
- Only two for Roman historian Tacitus.[9]

Normal L. Geisler reports, "In addition to this [the Greek manuscripts] there are 36,000 quotations of the New Testament by the early church fathers. Because of this, all but 11 verses could be accurately reconstructed even if we had no manuscript copies."[10] In addition, over 10,000 copies of the Latin Vulgate and at least 9,300 copies of early versions of the Bible in various languages are in existence. In total, over 24,000 manuscripts attest to the accurate transmission of the New Testament. No other ancient book can boast the same level of authenticity as the Biblc. It is distinct from all other books in message, unity, sophistication, and prophecy.

Summary of Bible Authenticity

Dr. Robert Wilson, a former professor at Princeton Theological Seminary, states: "After forty-five years of scholarly research in biblical textual studies and in language study, I have come now to the conviction that no man knows enough to assail the truthfulness of the Old Testament. Where there is sufficient documentary evidence to make an investigation, the statements of the Bible, in the original text, have stood the test."[11]

The Bible, the most ancient of preserved books, possesses many wonders. The thousands of existing manuscripts attest to God's preservation of its sacred content. The Bible's formation is unique, coming from some forty (mostly uneducated) writers and over a 1600 year period instead of by one religious founder. The full library is composed of sixty-six God inspired books which read as one. It is the best seller year after year, but also the most hated book. It is relished by the rich and the poor, by the scholar and the naïve.

The Uniformity of the Bible

Besides authenticity, another unique facet of the Bible which gives evidence of supernatural origin is its uniform content. The Bible was written over a 1,600 year period by some forty different writers who were situated in various social backgrounds and geographical locations, yet uniform truth is displayed throughout. Often these prophets of old did not fully understand the meaning of the very words they uttered on God's behalf (1 Pet. 1:10-12). This fact puts the Bible in stark contrast with the religious books of the world, which are generally composed by one individual – a religious founder.

How did God accomplish such uniformity? He controlled the speech and pens of the prophets; the Bible was literally God-breathed (2 Tim. 3:16). Scripture then, is a direct expression of both His truth and His love to mankind. Because God understands our natural limitations to comprehend spiritual and eternal matters, He utilized various literary forms in the Bible. These include word-pictures, poetry, prophecies, shadows, types, allegories, symbols, plain language, historical narratives, the gospels, and epistles. All of these speak of His supreme gift of love – His own Son to the world. Thus, the uniform focus of the Bible centers in the progressive revealing of God's purposes in history that culminate in Jesus' life, death, and resurrection.

The main theme of the entire Bible centers in Jesus Christ: *"For God so loved the world that He gave His only begotten Son, that whoever believes in Him [Jesus] should not perish but have everlasting life"* (John 3:16). *"For the testimony of Jesus is the spirit of prophecy"* (Rev. 19:10). *"For all the promises of God in Him [Christ] are Yes, and in Him Amen"* (2 Cor. 1:20).

To keep the plan of salvation in Christ a mystery until after His death and resurrection (1 Corinthians 2:7-8 states He would not have been crucified if the plan had been known), the messianic prophecies concerning Christ are generally scattered, seemingly sporadically, throughout the Old Testament. The truth was declared, yet in such a way that full understanding of the events and benefits of Calvary would not be understood from just one text. Furthermore, most of the Old Testament pictures of Christ are concealed in abstract symbols, reclusive personal portraits and mysterious names. Though these Old Testament gems were once concealed from human comprehension, they accentuate Christ when illuminated by the light of New Testament revelation.

As one investigates the Old Testament with the light of the New Testament, these abundant pictures and types of Christ are understood. Truly, the New is in the Old contained, but the Old is by the New explained. Volumes of books have been written on these striking preludes of realities to come. It is no exaggeration to state that hundreds of pictures and shadows of Christ are contained in the Old Testament; here are a couple notable ones:

The Ark

The ark bore God's wrath because of human sin, but all those (Noah and his family) who entered the ark by faith were saved from judgment (Gen. 6-8). The ark had only one door, which God personally opened and closed; this pictures the salvation and security that is in Christ. By water God judged the Earth, but the very water which destroyed life also lifted those in the ark off the Earth. In Christ, God separates a company of redeemed people from the world unto Himself.

The Passover Lamb – The Lamb of God

The Passover was instituted by God as a means of reminding the Israelites of their deliverance from slavery in Egypt and their restoration to God through redemption, which was obtained by personally applying the blood of an innocent substitute – a lamb. A male lamb without blemish in the prime of its life was to be tested for four days to prove its healthy condition. It was then killed, roasted, and eaten (any remnants were to be burned). The blood of this lamb was applied to the doorway of a household that desired to be spared the judgment of God on Egypt – the death of the firstborn (Ex. 12). The activity marked a clear distinction between the Egyptians and Israelites – applied blood on the door brought life, and no blood brought death.

John the Baptist declared that Christ was *"the Lamb of God which takes away the sin of the world"* (John 1:29). Paul taught that Christ was the literal fulfillment of the Passover Lamb: *"For indeed Christ, our Passover, was sacrificed for us"* (1 Cor. 5:7). Just as the Passover lamb was totally consumed or burnt, Christ was completely judged at Calvary for human sin. The millions of lambs previously slaughtered constantly reminded man of his sin, and testified that animal blood could not fully atone for his sin; it was necessary for the perfect, unblemished, fully-tested Man, the Lamb of God, to shed His blood. God sent His beloved Son to bear our sin (2 Cor. 5:21), to die (Heb. 2:9), and to redeem by His own blood anyone trusting in Him alone for salvation (1 Pet. 1:18-19).

The Uniformity of Content

Christianity is quite unique in that its *founder*, Jesus Christ, did not Himself personally write any of the Bible, whereas most world religions have a religious book which was written by a single individual claiming unique revelation. The Bible, however, was written by many writers from different cultural, geographic and social settings over a 1600 year period, yet the agreement of its sixty-six books is uniform. How is this possible, unless the Bible was orchestrated by one mind – God's. Here are a few examples of the Bible's uniformity.

Principle of First Mention

This principle in scriptural study implies that the first mention of a particular *key word* or phrase in the Bible usually establishes that word's general application throughout all of Scripture. The principle of first mention is witnessed in the life of Abraham in Genesis 15. The word of God came to Abraham, without any associated signs and wonders. God affirmed that, though Abraham didn't have any children, He would make him the father of many nations and, subsequently, bless all the Earth through him. The word of God was good enough for him – he simply trusted what God said and believed. God responded by accrediting a standing of righteousness to Abraham's account.

This accrediting, or accounting, of divine righteousness to a sinner exercising faith is seen throughout the Bible and is thoroughly explained by the Apostle Paul in Romans 4 and 5. Obviously, God wanted no confusion on this matter for the words "believe," "counted," and "righteousness" all occur for the first time in the Bible by one divine declaration in Genesis 15:6. This verse is contained three times in the New Testament (Rom. 4:3, Gal. 3:6, and Jas. 2:23) where it is more thoroughly explained.

Symbols

Besides the divine message contained in the normal narrative, God also uses numbers, icons, metals, colors, names, etc. to convey information. These more abstract forms of revelation do not substitute for or supplement the clear teaching of Scripture, but rather reiterate the obvious message of Scripture through metaphor. When God introduces an object, a number, a color, etc. in a metaphoric presentation, that symbolic meaning is held consistently throughout all sixty-six books of the Bible.

The consistent use of icons, numbers, analogies, names, first-mention occurrences, fulfilled prophetic types and shadows, plus the plain and consistent teachings of the Bible prove it to be the orchestrated genius of one Mind. The following list provides a few examples of numbers, colors, materials, animals, etc., which are commonly employed throughout the Bible (my book *In Search of God* explains each more fully).

Metals/Materials
Gold – Purity, Holiness, Righteousness.
Silver – Redemption.
Bronze – Judgment.
Wood – Humanity.

Animals/Man

Lion – King: Gospel of Matthew.
Ox – Servant: Gospel of Mark.
Man – Humanity: Gospel of Luke.
Eagle – Deity: Gospel of John.

Colors

White – Purity, Righteousness.
Red – War, Bloodshed.
Scarlet – Humility/Service.
Black – Death, Sin.
Blue – Heaven.
Purple – Royalty.

Miscellaneous

Fire – Judgment.
Sword/Knife – Judgment.
Stars – Angels or Messengers.
Lampstand – A Testimony.
Rainbow – God's Promises.
Eyes – Seeing.
Circle/Ring – Eternity or Security.
Mountain – Kingdom.
Horn – Power.
Head – Authority.
Dragon – Satan.
Earth – Israel.
Sea – The Nations.

Numbers

Each number, from one through forty, is used symbolically in the Bible; here are just a few examples.
One – Unity.
Four – Earthly Order.
Six – Man's number.
Seven – God's number (completeness and perfection).
Twelve – Governmental perfection.
Thirteen – Rebellion.
Twenty-four – The Priesthood.

Apparent Contradictions

Skeptics often use the argument that apparent contradictions within Scripture prove that the Bible is fallible and, therefore, not the Word of God. Sometimes this supposition is then used to prove that the God of the Bible Himself is fallible. While a few scribal errors will be found within the Bible as it exists today, no contradictions of context will be found in the Bible. If we think we have found a contradiction, it is because we have not properly discerned God's meaning of Scripture; *"For as the heavens are higher than the Earth, so are My ways higher than your ways, and My thoughts than your thoughts"* (Isa. 55:9). Man is to search all of Scripture for an understanding which is upheld by all of Scripture – the truth is in the whole.

Concerning actual scribal errors in the Bible the most noticeable are found in some Old Testament numbers. For example, about one sixth of the numbers in the parallel narratives of the Kings and Chronicles do not agree. C. I. Scofield explains:

> Some discrepant statements concerning numbers are found in the extant Hebrew manuscripts. Error by scribes in transcription of Hebrew numbers was easy, whereas preservation of numerical accuracy was difficult. Inspiration extends only to the inerrancy of the original autographs.[1]

This type of scribal error is generally easy to identify, and reason should guide the proper interpretation. For example, given Hebrew number construction, it would be logical to conclude that only seventy men from the small village of Bethshemesh died for the offense of peering into the Ark of the Covenant, in lieu of 50,070 (1 Sam. 6:19). Let us not forget that the Bible has been scientifically proven to be 99.5 percent authentic to the autographs and that the autographs were inspired by God – not translations of the autographs. Not understanding proper context, cynics have suggested many supposed biblical contradictions. Here are two examples with contextual explanation.

34

Time

Each Gospel presents a unique vantage point of Christ to a particular audience. The Gospel of Matthew presents Christ as the rightful heir to the throne of David and was primarily written to the Jews. Mark was written to the Romans (half of the empire were slaves) and conveys the lowly servant nature of Christ's ministry. Luke upholds Christ's humanity, thus identifying with a Greek audience. John, which presents Christ's deity, is addressed to the entire world. John refers to the world eighty times in his Gospel, compared to eighteen references in Matthew, five in Mark and ten in Luke. John has over twice as many references as the other three Gospels have combined.

Unlike the synoptic Gospels, John used the Roman versus Jewish reckoning of time. This is important to understand, otherwise, there would appear to be serious disagreement between the Gospel writers on major events in the Lord's life. For example, John records that Christ was in the judgment hall before Pilate at the sixth hour, but Matthew states that while Christ was on the cross, darkness covered the land at "the sixth hour." The sixth hour by Roman reckoning would be six o'clock, but the Jews would understand it to be twelve o'clock.

The Temptation (Testing) of Christ

If the accounts of Matthew 4 and Luke 4 are examined closely, one would notice that the order of Satan's specific attacks upon Christ are different. The skeptic deplores such inconsistency: "See, you can't trust the Bible – it does not agree with itself." However, the order maintained by each writer is for the purpose of upholding the prevalent theme of each Gospel. In Matthew, Satan first asks the Lord Jesus to turn the stones into bread, then bids Him to cast Himself down from the pinnacle of the temple, and thirdly offers to Christ all the kingdoms of this world if Christ will only worship him. Luke's order, however, is the request to turn the stones into bread first, then the offer of the kingdoms of the world, and finally Satan adjures Christ to cast Himself down from the pinnacle of the temple to prove that the angels will

35

protect Him. So why do Matthew and Luke portray a different order of events? Because Matthew is presenting Jesus Christ to a Jewish audience as the rightful heir to the throne of David, while Luke's upholds Christ's humanity and thus records the events as each occurred. It is sovereign design which accounts for the variation of the accounts: Luke's order of temptations is chronological, while Matthew's is climactic unto kingship.

Understanding Context

It is usually people who know nothing or little about the Bible who identify biblical contradictions. Most supposed contradictions result from poor contextual insight or not differentiating the ongoing economies of truth God reveals throughout the Bible.

The skeptic complains, "The Bible teaches 'Eye for an eye and tooth for a tooth,' but in another place the Bible says, 'pray for your enemies and do good to them who persecute you.' Therefore, the Bible obviously disagrees with itself." The former truth was expressed to the Jews to teach them that disobedience must be punished if the Law was not kept. Whereas the latter statement reflects a deeper truth pertaining to the New Testament; Christians had already realized that no one could earn heaven by keeping the Law and had trusted Christ as Savior. Having received the Holy Spirit, Christians had the wherewithal to both *keep* the Law (i.e. choose not to sin), and *fulfill* the Law (demonstrate God's gracious character, see Rom. 13:8-10).

Throughout the human timeline, God has diligently worked to cause man to be more aware of his sin and of the divine solution for it – Christ. It is not that God is changing absolute truth, rather He is methodically unfolding it for man's benefit. The purpose of the Law was to make the Jews conscious of their sin and their personal need for a Savior (Rom. 3:20; Gal. 3:24).

Many of the "contradictions" in the Bible posed by skeptics are merely the picking and choosing of random statements which have been drawn out of their proper context and meaning, as determined by the whole of Scripture. The Bible is incredibly uniform and supposed contradictions are just not there!

Prophecy, the Proof of Inspiration

We have reviewed the Bible's authenticity and uniformity, now we focus on its third unique quality; its prophetic content. There are many unique qualities of the Bible, but these three (authenticity, uniformity, and prophecy), are the most prominent aspects which demonstrate supernatural origin.

The Bible repeatedly proves its validity by accurately foretelling thousands of future events; many of which are now verifiable history. This same prophetic content is not apparent in the Hindu Vedas, the Quran, the sayings of Buddha or Confucius, the scriptures of Shintoism, the Book of Mormon or any other religious writings. Approximately one fourth of the Bible is prophetic in content; in this distinction, it is in a class by itself as compared to any other "religious book."

Through prophetic statements, God puts His name on the line again and again to show the world that He is the one true God and that the Bible is His message to humanity. Ignorance is not bliss; man is without excuse! Irwin H. Linton says it well in his book, *A Lawyer Examines the Bible*: "To doubt is not sin, but to be contented to remain in doubt when God has provided 'many infallible proofs to cure it, is."[1]

A Prophetic Portrait of Messiah

Many irrefutable evidences prove that the Bible is God's truth, but none is more convincing than fulfilled prophecy. Much of the Bible's prophecy has to do with the nation of Israel, but the most significant are the hundreds of prophecies pertaining to the coming of the Jewish Messiah. Over two hundred prophecies pertain to Christ's first advent to the Earth and twice that number for His second return to rule and reign. As the second advent of Christ has yet to occur, those related prophecies are unverifiable. First advent prophecies have been fulfilled and may be analyzed.

We all enjoy looking at family pictures and reminiscing of bygone days. Though our outward appearance changes as we age, we are still able to identify certain characteristic features of loved ones in photographs and video clips. Through prophetic Scripture, God has likewise brought various writers together, over a huge span of time, to create a timeless prophetic portrait of the Messiah. They didn't use paint, but each one dabbed a bit of prophetic color in words into the canvas of Holy Scripture to depict Christ. Those willing to gaze upon the prophetic portrait embedded in Scripture would recognize Messiah's features when He arrived. In this way, God removed all doubt as to who the Messiah actually would be. The one claiming to be Messiah would indeed be Messiah, proving *every* detail in the prophetic portrait was fulfilled.

The prophetic hues of God's portrait of Messiah include over 200 Old Testament prophecies by more than a dozen prophets some 1,600 to 400 years prior to Christ's first coming to the Earth. After removing redundant prophecies and the more vague ones (which the skeptic might argue against) from consideration, I found sixty-nine major unquestionably Messianic prophecies in the Old Testament. Just as a picture confirms that no two people are exactly alike, no one could possibly fulfill all sixty-nine of these prophecies but Messiah Himself.

The following table contains a sampling of these biblical prophecies. A statistical evaluation of all sixty-nine prophecies is contained in my book *In Search of God*. The combined probability that the Lord Jesus Christ fulfilled all sixty-nine major prophecies, by chance, is estimated to be one chance in 5.32×10^{72} attempts. Prophecies beyond statistical possibility (Christ's virgin birth and His resurrection, for example) were given a mere 50/50 chance of fulfillment (i.e. He was either born of a virgin or not).

OT Messianic Prophecies	OT Reference/NT Record
Of the tribe of Judah	Gen. 49:10/ Rev. 5:5
No bone broken	Ex. 12:46; Ps. 34:20/ John 19:36
Cursed on the tree	Deut. 21:23/ Gal. 3:13
Seed of David	Ps. 89:3-4, 19, 27-29/ Matt. 1:1
Jews and Gentiles plot together	Ps. 2:1-2/ Acts 4:25-28
Hands and feet pierced	Ps. 22:1-31/ Matt. 27:31, 35
Time of darkness at death	Ps. 22:2; Amos 8:9/ Matt. 27:45
Mocked and insulted	Ps. 22:7-8/ Matt. 27:39-43, 45
Stripped of clothes	Ps. 22:18/ Luke 23:34
Soldiers cast lots for outer coat	Ps. 22:18/ Matt. 27:35/John 19:24
Soldiers divided inner garment	Ps. 22:18/ Matt. 27:35/John 19:23
Accused by false witnesses	Ps. 27:12; 35:11/ Mark 14:57-58
Betrayed by a friend	Ps. 41:9, 55:12-14/Mark 14:17-21
Betrayer would die without repenting to God	Ps. 55:13-15/ Matt. 27:3-5; Acts 1:16-19
Given gall (a narcotic) to taste	Ps. 69:21/ Matt. 27:34
Given vinegar to drink	Ps. 69:21/ Matt. 27:48
Would have a forerunner to make ready the way of the Lord	Isa. 40:3-5; Mal. 3:1/ Matt. 3:3; 11:10-14; Mark 1:2-3;
Be scourged	Isa. 50:6/ Matt. 27:26;
Be beaten, struck in the face	Isa. 50:6; Mic. 5:1/ Matt. 26:67
Spat upon	Isa. 50:6/ Matt. 26:67; 27:30
Silent when accused	Isa. 53:7/ Mark 14:61; John 19:9
Buried with the rich	Isa. 53:9/ Matt. 27:57-60
Would live in Galilee	Isa. 9:1/ Matt. 2:22
Journey into Egypt	Hos. 11:1/ Matt. 2:15
Born in Bethlehem of Ephrathah	Micah 5:1-5/ Luke 2:4, 10-11
He would visit the 2nd temple	Hag. 2:6-9/ Luke 2:27-32
Triumphal entry into Jerusalem on a colt, the foal of a donkey	Zech. 9:9-10/ Matt. 21:4-5; Mark 11:2-10; Luke 19:35
Sold for thirty pieces of silver	Zech. 11:12-13/ Matt. 26:14-15
Betrayal money returned to temple	Zech. 11:12/ Matt. 27:3-5
Betrayal money used to buy potter's field	Zech. 11:12-13/ Matt. 27:9-10
Piercing of His body	Zech. 12:10/ John 19:34, 37

So just how big of a number is 5.32 x 10^{72} (that is 532 with 70 zeros behind it)? What would be the probability of someone randomly choosing a previously marked grain of sand from a child's sandbox? The probability would be extremely unlikely. What would be the chance of randomly choosing one particular grain of sand out of a sandbox the size of a county, or a country, or a continent? The odds would be astronomically impossible. Now let us suppose that we could convert the entire mass of the universe into grains of sand and that one single grain of sand could be marked and hidden anywhere in the universe (which, as stated earlier, some scientists say is nearly 100 billion light years across). What would be the probability of randomly picking out the previously marked grain of sand by chance?

The mass of the universe is estimated to be 1.6 x 10^{60} kg, while the measured mass of a 2 mm grain of sand is about 9.0 x 10^{-5} kg.[6] The resulting probability of picking the previously marked grain of sand from somewhere in our galactic sandbox is one chance in 1.78 x 10^{64} tries. This is beyond astronomically impossible! But as improbable as this seems, the Lord Jesus fulfilling all sixty-nine major Old Testament prophecies is 100 million times *less* likely than randomly choosing that one grain of sand from any location in the universe!

The probability of Christ fulfilling all the Old Testament prophecies by chance is absolutely quantifiably impossible – He fulfilled each one by design. The prophetic portrait of the Messiah painted centuries earlier on the Holy Page proves beyond any shadow of doubt that Jesus Christ is who He said He was – He is God's Messiah. Some might say, Jesus knew the Old Testament Scripture, and He was trying to fulfill each one by human effort. How is this possible, as the vast majority would not be within the control of human effort. How does one control his or her genealogy, time of birth, birth place, childhood travels, means of personal torture, the method of one's death, and his or her burial place? Only God could control such details!

Some of the scribes and religious leaders of Christ's time memorized large portions of Scripture. Yet, knowing Scripture, they still paid Judas the betrayal money in silver, not gold, and it was the exact amount that Zechariah had prophesied over 500 years earlier. Christ was not stoned, the normal means of execution under Jewish law, but crucified. This fulfilled what David wrote in Psalm 22, a thousand years earlier, that Messiah's hands and feet would be pierced. If these religious leaders had been able to void one prophecy, it would have proven that Jesus Christ was not the Messiah; instead, their religious blindness made them available instruments in the hand of a sovereign God. God proved to the world that His Son was the long awaited Messiah.

Summary

Nostradamus published his book of prophecies over four centuries ago, but their vague content has allowed various interpretations to fit a number of historical situations. Accordingly, most of his prophecies are unverifiable. For those prophecies which may have been fulfilled, this is really of no concern, for Satan himself (allowed by God) is able to induce great wonders to deceive man, thus providing a test of man's resolve to adhere to God's Word (2 Cor. 11:13-15). The test of divine prophecy is not that many future proclamations of a particular prophet come true, but that one hundred percent of what was prophesied actually happens. Anything less is not of God (Deut. 13:1-5).

As stated earlier approximately one-fourth of the Bible is prophetic in nature. Most world religions, such as Islam, Buddhism, and Hinduism shun specific prophetic utterances. Why? Unfulfilled prophecies and wrongly predicted events provide evidence of deceit and not of divine truth. Consequently, the failed prophecies of Joseph Smith and Brigham Young (founders and so-called prophets of Mormonism) or the numerous false prophecies of Charles Russell (the founder of The Jehovah's Witnesses) serve as examples of what is not of God. My book, *Hiding God*, documents dozens of failed prophecies by the founders of these two false religions.

In contrast, the Bible is full of specific and verifiable prophecies. In this regard the Bible is unique; no other secular or religious prophet can compete with the Bible's prophetic content or clarity. As further example, the prophet Isaiah named Cyrus, a future Persian king, two centuries prior to his birth. Cyrus was to be, and indeed was, God's instrument to topple the Babylonian empire, release the Jews from captivity, and rebuild the temple in Jerusalem (Isa. 44:28-45:1; Ezra 1:1-2). And incredibly, neither the Babylonian, nor the Persian empires existed at the time of this prophecy.

The Bible's evident authenticity, uniformity, and demonstrated prophetic content offer astounding proof of its supernatural origin. Scientific and historical evidence verify the Bible to be true! The Bible has withstood thousands of failed attempts to destroy its truth and to discredit its divine authority. God has accurately preserved His Word so that man might know Him and understand what is required of him. Despite the centuries of relentless attack, the Bible's truth and authority stand strong. "Holy books" contrived by world religion cannot survive the scrutiny of external and internal examination; only what is true will stand the test of time and the Bible is still standing.

The Challenge

If you have not explored the Bible, I challenge you to read it and to try to avoid being brought under its influence. Many profound atheists have sought to disprove the validity of the Bible and have become convinced of its message. Some who once rejected its message, such as C. S. Lewis, Josh McDowell, and Lee Strobel, have since written their own defenses of the Christian faith.

After experiencing the transforming power of the Bible, C. S. Lewis wrote: "A young man who wishes to remain a sound atheist cannot be too careful of his reading. There are traps everywhere – Bibles laid open, millions of surprises."[2] If you are not consistently reading the Bible, why not commit some time to reading it and mediating on what you have read? The Bible's influence is amazing. I recommend starting in the Gospel of John.

Worldviews, Religion, and Revelation

The term *worldview* refers to any ideology, philosophy, theology, etc. which provides an intellectual framework for interpreting life, including the metaphysical properties associated with existing. Volumes have been written on this subject, it suffices here to merely introduce and briefly comment on the philosophical integrity of each major worldview. The seven major worldviews are as follows:

- Atheism: no God (time + random accidents = reality).
- Polytheism: many gods (random acts, limited control).
- Pantheism: God *as* the universe = the impersonal One.
- Panentheism: God *in* the universe (his body), but he also has a spiritual existence beyond nature (a soul).
- Finite Godism: God is limited in goodness and power; cannot counter man's wrong doings.
- Deism: One God initially created all, but he is not in control now – initial laws working to a planned end.
- Theism: One personal all powerful God who created all, continues to control all, and is distinct from all.

The atheistic worldview declares that there is nothing but nature. Closely associated with the atheist is the agnostic, who is pacified not knowing if there is anything beyond nature; certainly nothing more than nature is needed to make sense of life.

The supernaturalist believes that there was an initial and personal cause to create nature and an ongoing influence, or at least a predetermined law, to maintain it. This fundamental perspective is upheld by the remaining major worldviews (theism, deism, pantheism, polytheism, finite godism, and panentheism).

Logically Evaluating Worldviews

At least two conclusions are inescapable concerning the validity of worldviews: First, because each worldview contradicts all others, at most, one can be correct. Secondly, for a worldview to be correct it must be logically consistent in principle. When critically evaluated only theism is shown to be non-contradictory in principle.

Atheism, for example, cannot define a first cause and thus contradicts its most fundamental principle that all beliefs must be supported by observational evidence and those which cannot must be rejected. Of course rationally speaking, only an all-knowing, unchanging, eternal being would be able to categorically declare "There is no God," but then would not such a being actually be God? In either case, atheism is proven to be an illogical assertion.

The worldviews of polytheism, pantheism, panentheism, and finite godism identify a supernatural presence which is limited, or initiated or changeable. Being that is subject to natural laws (e.g. cause and effect) is contingent being (i.e. existence resulting from something which previously existed). Contingent beings therefore cannot account for their own existence; a superior necessary being must exist. For example, humans are contingent beings – we exist because our parents existed; Adam and Eve existed because God created them. Through procreation (the natural law of reproduction) man simply passes along to the next generation the building blocks of life that previously existed. The changeable or limited god(s) of these worldviews have a contingent essence which requires a pre-existing being to account for their existence.

Deism contradicts itself in principle by stating a self-existing, unchanging, unlimited God created all things, but then cannot intervene to manipulate that which He first created. Certainly any subsequent divine manipulation of nature would be an inferior feat in comparison to the initial miracle of creation. Philosophically speaking, theism is the only worldview which does not contradict itself – one all-knowing, all-powerful, eternal, immutable God created all and controls all.

Belief-systems relating to supernatural worldviews are numerous. So which, if any, is based on divine truth and which is merely a religious facade? As with worldviews, at least two conclusions pertaining to belief-systems are apparent: First, only one of these belief-systems can represent truth, as the tenets of each system are incompatible with others. Secondly, for a belief-system (such as a particular religion) to be correct, the basis of that system, such as "divine revelation" usually in the form of some type of "holy book" must be consistently accurate. Why? Because absolute truth cannot be inconsistent.

So which belief-system can boast of a foundational truth which is both consistent in content and agrees with scientific discovery? I believe, as shown in earlier chapters, the only answer is a belief-system which accurately adheres to the Bible's teachings. This eliminates "Churchianity" and religiosity in Christian disguise; biblical Christianity is not to be confused with these humanized forms. No denomination, no sect, or no movement which systemizes Scripture can claim to be perfectly true. The truth of the Bible is in its entirety, and man, even when he tries his best, will fall short of obeying the precepts of Scripture.

With this said, it is understood that there is one central truth concerning salvation through the Lord Jesus Christ that permeates the entire Bible. When an individual yields to this revelation, he or she is brought into a relationship with God. Thus, *biblical Christianity* is not an earthly institution, or a human-conspired organization, or a set of systematized teachings per se, but rather it is the opportunity to experience spiritual union with Jesus Christ – this is the central theme of Scripture. Oneness with Christ secures eternal life and affords the believer an opportunity to enjoy a life that is meaningful and pleasing to God (Luke 9:23-26).

The main distinction between biblical Christianity and all the religions of the world is that it teaches a vital need to be saved from spiritual death by trusting in a Savior alone, whereas the world's religions present a system of *doings* to merit salvation or to obtain an improved afterlife. Religion equips man with a "do it yourself" manual and workbook through which he may impress himself as to how well he is *doing* by completing religious exercises/checklists. Christianity, however, is not a *religion*; it is a *relationship* with Jesus Christ. Apart from Christ, there is no forgiveness of sins, no life, and no hope. This is the message in the Bible: Jesus said, *"I tell you, no; but unless you repent you will all likewise perish"* (Luke 13:3). *"Jesus said to him, 'I am the way, **the truth**, and the life. No one comes to the Father except through Me'"* (John 14:6).

World Religion and New Revelation

A common methodology among theistic religions developing after the divine issuance of the Bible, is to insist that new revelation from God is warranted because the message of the Bible is no longer authentic – they claim that the Bible has been corrupted and its message can no longer be trusted. The end goal is that biblical truth must be concealed, altered, or at least significantly deemphasized. Examples would be Muhammad's Quran, Joseph Smith's Book of Mormon (with *Doctrines and Covenants*), the Jehovah's Witnesses' New World Translation, and the Roman Catholic Catechism.

The "holy books" of Islam and Mormonism lift various stories, teachings, verses, phrases, names, etc. from the Bible to confuse and blur the truth. Roman Catholicism is progressively replacing the Bible with Church Tradition, and the Jehovah's Witnesses distort the truth of Scripture through Watchtower interpretations of a tainted translation of the Bible. Deformation of the truth, often in progressive degrees, is a frequent tactic of Satan to lure men from divine truth. Here are some examples.

Islam

The *Quran* did not exist at the time that Muhammad died. Based on Islamic tradition it is believed that Muhammad could neither read nor write (Quran 29:48) but that he recited what was revealed to him for his companions to write down and memorize. After Muhammad's death in 632 A.D., numerous versions of his teachings were in circulation. Some twenty years later, Uthman (the third Caliph of Islam according to Sunni Muslims) ordered that a collection be made of available writings and that these be put into some order. Imperfect men do not memorize or write perfectly, so one can only imagine the huge number of discrepancies between all the pieces of composition collected.

If Muhammad was truly the last prophet of God, then the vast differing assortment of Muhammad's teachings were pieced together by various non-prophets (non-inspired individuals) to create the Quran. After the collection was compiled, reduced and ordered in 656 A.D., all other versions were to be burned. The final version of the Quran had 114 "Surahs" or chapters. The fact that Muslims, themselves, have difficulty explaining how the Quran was actually assembled, and that there were many early forgeries cast doubt on the integrity of the book as representing exactly what Muhammad said. Sharon Morad Leeds summarizing the book *The Origins of the Quran: Classic Essays on Islam's Holy Book* writes: "Modern Muslims assert that the current Quran is identical to that recited by Muhammad. But earlier Muslims were more flexible. 'Uthman, A'isha, and Ibn Ka'b (among others) all insisted that much of the Quran had been lost"[1]

There are many contradictions within Quran. Examples: Is there or is there not compulsion in religion? The Quran contradicts itself (2:256, 9:3, 5, 29). Was the first Muslim Muhammad (39:12) or Abraham (2:132) or Jacob (2:132) or Moses (7:143)? Does Allah forgive or not forgive those who worship false gods? The Quran disagrees (4:48, 4:116, 4:153). From what material did Allah create man from? "A [mere] clot of congealed blood" (96:2). "From clay, from mud molded into shape" (15:26). "From dust" (3:59). "Out of nothing" (19:67; 52:35). "From a sperm-drop" (16:4). Obviously, all these cannot be true.

Mormonism

If you are familiar with the Bible and were to read the *Book of Mormon*, you would quickly realize that extensive portions were plagiarized by Joseph Smith from the King James Version of the Bible. For example, Moroni 10:8-17 is copied from 1 Corinthians 12:4-11 and 2 Nephi chapters 12 through 24 are copied from Isaiah chapters 2 through 14. This means that some portions of the *Book of Mormon* were drawn from Scripture, while other portions were humanly improvised.

It is the non-scriptural portions which can be easily proven not to be of divine origin. Truth cannot be modified and still remain error free, consequently leaders in the Mormon Church have significantly revised the *Book of Mormon* by introducing new material and in many cases falsely attributing it to an earlier date. The *Book of Mormon* is less than 200 years old, so why have there been over 3,900 changes to it, if in fact it is, as Joseph Smith declared, "the most correct of any book on Earth, and the keystone of our religion, and a man would get nearer to God by abiding by its precepts, than by any other book."[2] The book was first written in English by Joseph Smith, so these changes are not translational refinements of Greek or Hebrew texts, but wholesale change! Divine truth does not need to be corrected – it is timeless.

How has science confirmed the truth of the *Book of Mormon* (speaking of the non-Bible copied portions)? Hal Hougey explains, despite years of work by archaeologists the following has become clear concerning the historical record of events, places and people in the Americas found in *The Book of Mormon:*

1. No 'Book of Mormon' cities have been located.
2. No 'Book of Mormon' names have been found in New World inscriptions.
3. No genuine inscriptions have been found in Hebrew.
4. No genuine inscriptions have been found in Egyptian or anything similar to Egyptian, which could correspond to Joseph Smith's "Reformed Egyptian."
5. No ancient copies of 'Book of Mormon' scriptures have been found.
6. No ancient inscriptions of any kind which indicate that the ancient inhabitants had Hebrew or Christian beliefs - all are pagan.
7. No mention of 'Book of Mormon' persons, nations, or places have been found.
8. No artifact of any kind which demonstrates the 'Book of Mormon' is true has been found.[3]

Archeological and other scientific evidence (such as DNA analysis)[4] contradicts *The Book of Mormon*'s accounts of how the American Indians originated, the earliest explorations of the New World, plant and animal life in the Americas, use of metal, steel, and silk by ancient Americans; and other supposed similarities between pre-Columbian Indian cultures and those of the Old World.[5] The absence of any corroborative evidence from external sources and the constant need to revise the Book of Mormon to fix doctrinal problems proves that it is not divinely originated.

Jehovah's Witnesses

For nearly seventy years, the Jehovah's Witnesses relied on the King James Version or the American Standard Version (1901) of the Bible in their organizational writings. As developed Watchtower doctrines could not be reconciled with these Bible translations, an effort was launched in the late 1940s to create one that would. The result was the *New World Translation*. The New Testament was available in 1950 and the Old Testament in 1961.

Interestingly, only six individuals served on the New World Translation committee (this is a very small committee compared to the committees involved with other translations of the Bible), and only one committee member, Frederick William Franz, had any formal training in biblical languages.[6] His training consisted of 21 credit hours of *Classical* Greek including some Latin, and only two credit hours of *Koine* Greek, the language of the New Testament. To summarize, Franz had only two formal credit hours of training in the language of the New Testament and could not read Hebrew, the language of nearly all the Old Testament (this was proven in the court case of Scotland - Douglas Walsh vs. The Right Honorable James Latham Clyde – November 1954).[7]

The New World Translation consists of thousands of changes to the Bible (most of them subtle) to advocate Jehovah's Witness teachings. Jehovah's Witnesses will commonly say that they derive their doctrines from the Bible only. However, the organization's regular publication *Watchtower* expresses its authority to define what is to be derived as doctrine from the Bible; for only the organization can understand and properly interpret the Bible (e.g. Watchtower; Dec. 12, 1990, p. 19).

Starting with Charles T. Russell, then Joseph F. Rutherford and Fredrick W. Franz, their history is plagued with false prophecies! It is noted that some Internet websites list over one hundred such false prophecies of the Watchtower Society. For example, early Watchtower prophets foretold Christ's second coming would be on certain dates (e.g.,1874, 1914, and 1975).[8] After each erroneous prophecy a well-executed cover up, that spiritualized the meaning of the prophecies, was developed to minimize organizational damage.

Buddhism

Some 2,500 years ago, a man from India named Siddhartha Gautama (Buddha), supposedly discovered the cause of unhappiness and its cure after sitting under a tree for forty-nine days. From what he gained during this enlightenment period he spent the rest of his life teaching others. Buddha, an atheist, never intended to start a new religion; rather his goal was to cleanse Hinduism from its idolatry. Ironically, the idolatry that Buddha condemned is now a major part of Buddhism.

Buddha wrote down none of his teachings; in fact, the revelation was orally transmitted for centuries. By the time Buddhists did record his teachings, there was a lack of consensus as to what those teachings were. Consequently, the teachings of Buddhism, as originated by Buddha, cannot be substantiated.

Summary

Should religious "holy books" be trusted as a guide to finding the truth? The answer is "no, what is proven to be false should not be trusted as divine truth"! What is truly from God will be consistently true and what is not is counterfeit revelation. Divine revelation will be found accurate, uniform, and, if prophetic, will occur. What is truly God's written Word will have proven authenticity, and will be in agreement with accurate external evidence. Only the Bible can satisfy this criterion.

Your Spiritual Journey

Before embarking on a long journey, it is wise to obtain a road map to ensure you will find the desired destination. Life itself is a long spiritual journey, and likewise we need a map of the way. But which spiritual map is to be trusted? Many spend significant time planning a family vacation or a sightseeing trip, but neglect to properly prepare to venture into eternity.

How are we to know the truth concerning eternity? How can we discern if a "Holy Book" is divinely inspired or humanized religion? The answer to the latter question supplies the answer to the former question. As previously shown, I believe the Bible is God's divine map to guide man heavenward because the overwhelming internal and external evidence of its truthfulness.

Naturally speaking, if we were left to ourselves, we would never know God personally because we would never know how to seek after Him. After our first parents sinned, spiritual death (separation from God) passed down to all their descendants (Rom. 5:12). That is why Scripture states that an individual is already condemned even before he or she might reject the gospel message of Jesus Christ (John 3:18). Paul summarizes our natural spiritual condition, *"There is none righteous, no, not one; There is none who understands; There is none who seeks after God. They have all turned aside; They have together become unprofitable; There is none who does good, no, not one"* (Rom. 3:10-12). Man is hopelessly lost and separated from God as a result of sin.

The Matter of Sin

Sin is not a popular word. It describes "lawlessness," "rebellion," "not doing what we know is right," "falling short of God's standard of righteousness" (Rom. 3:23; James 4:17). God's moral standard of right and wrong is declared to man in the Ten Commandments. These commandments (the Law) show us our sin (Rom. 3:20) and affirm that we need a Savior (Gal. 3:24-25).

The first two of the Ten Commandments relate to the subject of recognizing God as Creator and not worshipping creation: *"You shall have no other gods before Me"* (Ex. 20:3). *"You shall not make for yourself any carved image"* [something worshipped or adored more than God] (Ex. 20:4). These commandments alone are sufficient to prove that each of us has sinned against God, but in case you are not convinced, here is a paraphrased summary of the remaining commandments (Ex. 20:3-17):

- Do not blaspheme God or use His name disrespectfully;
- Put aside one day in seven to honor the Lord;
- Honor your parents;
- Do not murder;
- Do not commit adultery;
- Do not steal;
- Do not lie;
- Do not covet (lust after what is not yours).

Ask people on the street if they are a good person, and most will say "Yes, I am a pretty good person." Their moral standard of reckoning, however, is all wrong and they don't even know it. They have fabricated a self-righteous system of weighing their good deeds against their bad (sin), thinking that their good deeds will somehow offset their sins. God's standard of judgment is quite different – absolute perfection! By His standard, one sin will keep anyone out of heaven (Gal. 3:10-12). Someday each of us will be judged by God's standard of perfection (Eccl. 12:14; Rom. 14:10-12; Rev. 20:11-15). How do you measure up against His standard of perfect holiness?

The Bible answers that question for us on two counts. First, as to the matter of sin: *"All have sinned and fall short of the glory of God"* (Rom. 3:23), and secondly as to good works: *"All our righteousnesses are like filthy rags"* (Isa 64:6). So, one sin will tip God's moral scales against us because good works cannot satisfy God's righteous demand that our sins must be punished.

Imagine for a moment that you are driving a car faster than the posted speed limit and, appropriately, are pulled over by a highway patrolman. While he is writing you a citation for speeding, you boast to the patrolman that you are a good parent and a good spouse, that you do community service work, that you give generously to charities, etc. To your frustration, the patrolman continues writing and hands you the citation – you are fined $250. What is the moral of the story? No amount of good works undoes the fact that we all have violated God's moral law. Now ponder all the self-righteous and religious fabrications that men have concocted to convince God that they do not deserve judgment: water baptism, good works, religious parents, church attendance, repetitious prayers, rubbing beads, reciting religious chants, etc.

In our natural state apart from God, our good works do absolutely nothing to earn God's favor. The flesh nature opposes God in deed and motive; therefore, there is nothing inherent within the flesh that can please God. Jeremiah states: *"The heart [seat of emotion] is deceitful above all things, and desperately wicked"* (Jer. 17:9). The grand conclusion is that we all are sinners and that we can do nothing to persuade God to love us more than He already does. Good works are evidence of true salvation in Christ, not the basis for salvation (Jas. 2:17, 20).

The Consequence of Sin

How then can man obtain a pure standing with God? The answer is that each individual must be justified (declared right by God). This is how Abraham received salvation: *"Abraham believed God, and it was accounted to him for righteousness"* (Rom. 4:3). If anyone could have merited salvation through personal effort, it would have been Abraham, but Scripture condemns him apart from receiving the righteousness of God. Justification results when divine rightness is imputed or accredited to one's account. When an individual trusts Christ, God imputes a righteous standing to that individual's account (he or she is declared righteous before God, though in practice he or she will still sin). This reality is a positional truth which the Christian is to practically live out daily (Rom. 6:11-12, 13:14).

If you die without being justified (i.e. receiving forgiveness of your sins and obtaining a righteous standing in Christ), there is no hope for you (Heb. 9:27). Contrary to what some teach, there is no purgatory – you will spend eternity in hell. God has done everything He can to rescue you from eternal judgment, but He will not force you to go to heaven – it is your choice (2 Pet. 3:9). The Bible vividly describes hell, the ultimate fate of those who reject God's truth:

- *"Shame and everlasting contempt"* (Dan. 12:2)
- *"Everlasting punishment"* (Matt. 25:46)
- *"Weeping and gnashing of teeth"* (Matt. 24:51)
- *"Unquenchable fire"* (Luke 3:17)
- *"Indignation and wrath, tribulation and anguish"* (Rom. 2:8-9)
- *"Their worm does not die* [putrid endless agony]*"* (Mark 9:44)
- *"Everlasting destruction"* (2 Thess. 1:9)
- *"Eternal fire ... the blackness of darkness for ever"* (Jude 7, 13)
- *"Fire is not quenched"* (Mark 9:46)

Revelation 14:10-11 tells us the final, eternal destiny of the sinner: *"He shall be tormented with fire and brimstone ... the smoke of their torment ascended up for ever and ever: and they have no rest day or night."* The Bible's teaching of an eternal place of punishment for unforgiven sinners offends people; consequently, many are watering down the truth – teaching that hell is a state of non-existence or quick annihilation. Misrepresenting the truth to avoid its consequence is never a good idea.

The Lord Jesus spoke more about hell than He did heaven, addressing the subject over seventy times. He bore our hell at Calvary and appeased God's anger for our sin so that we would not have to spend eternity there. He did not frighten people for kicks. He lovingly warned them about their deadly disease (sin) and the fatal consequence of the disease (hell), and pleaded with them to internalize the cure (exercise faith in Him for salvation).

God will not force anyone against his or her will to receive Christ so that they might live with Him in heaven. Heaven would be hell if you didn't want to be there, but hell will not be heaven for those rejecting Christ. Salvation is like a personal check that has been written out for the full value of our offenses against God which God extends to us in the person of Christ. As an individual believes on Christ, he or she is by faith, endorsing the check and the value of it is imputed to their personal account. The check has value whether we cash it or not, but it only has value in our account if by faith we take action to cash it. God will not force your signature – the choice is yours.

The Solution to Sin

The Lord Jesus said, *"Unless you repent you will all likewise perish"* (Luke 13:3). Repentance means that you agree with God that you are a sinner deserving His judgment, and that you first turn away from all you thought would earn you heaven; such repentance indicates a deep grief over personal sin and a desire to turn from wickedness (Jer. 8:6).

Secondly, you must turn to something – that is you must believe the gospel of Jesus Christ. To alleviate any confusion about what this message is, the Lord Jesus Christ personally conveyed it to Paul: *"Christ died for our sins according to the Scriptures, and that He was buried, and that He rose again the third day according to the Scriptures"* (1 Cor. 15:3-4). Believing any other gospel will result in eternal damnation (Gal. 1:6-9).

So if by faith one believes and receives Christ for the forgiveness of his or her sins, he or she is then spiritually born again (John 3:3; 1 Pet. 1:23) and becomes an adopted child of God (Rom. 8:15-16). Birth and adoption are acts which establish relationship; man's fellowship with God depends upon righteous behavior (1 Jn. 1:5-10). Ongoing godly behavior is now possible through the abiding presence of the Holy Spirit; if the believer chooses to obey revealed truth, his or her fellowship with God is unbroken. If one sins, fellowship is broken, but relationship is secured in Christ. As Christians sincerely confess their sins they are restored to full communion with God (1 Jn. 1:9).

The Lord Jesus said the way to eternal life is found only in the narrow way (Himself), but the path to destruction is wide – there are many ways leading into hell (Matt. 7:13-14). The Lord went on to say only a few would find the narrow way and the narrow gate which leads into the eternal bliss of heaven.

When we enter the narrow way by trusting the gospel message of Jesus Christ, God rewards us with Christ and all the riches that are in Him (Eph. 1:3). By God's mercy, the believer escapes hell, and by His grace, he or she inherits heaven and all that Christ has (Rev. 21:7). Will you not trust Christ for salvation and know the wonder of God? The Lord Jesus is both the beginning and the end of man's quest to uncover and believe the Truth! If you hear His pleading voice, please don't harden your heart; trust the Lord Jesus Christ for the salvation of your soul.

For God so loved the world that He gave His only begotten Son, that whoever believes in Him should not perish but have everlasting life. For God did not send His Son into the world to condemn the world, but that the world through Him might be saved. He who believes in Him is not condemned; but he who does not believe is condemned already, because he has not believed in the name of the only begotten Son of God (John 3:16-18).

Concerning one's spiritual journey into eternity, individuals have limited choices: to ignore the supernatural altogether, to have a "make me feel good" rapport with religion; or to have an intimate relationship with God through the Lord Jesus Christ. The reader must decide what is the more pertinent, reliable, and genuine path in life to venture down. If you decide that knowing God is more crucial than an intellectual link with science or the "feel good" remedy of religion, you must approach God through Christ alone; there is no other way to be saved. Speaking of Jesus Christ, Peter said: *"Nor is there salvation in any other, for there is no other name under heaven given among men by which we must be saved"* (Acts 4:12). How will you venture into eternity?

Endnotes

Truth Doesn't Argue with Itself
1. Edythe Draper, *Draper's Quotations from the Christian World* (Tyndale House Publishers Inc., Wheaton, IL), # 9867
2. William MacDonald, *Believer's Bible Commentary* (Thomas Nelson Publishers, Nashville, TN; 1989), p. 2103

Archeology Agrees with the Bible
1. Millar Burrows, *What Mean These Stones?* (Meridian Books, New York, NY; 1965), p. 42
2. Nigel Reynolds, "Tiny Tablet provides proof for Old Testament," (Telegraph.co.uk; July 13, 2007). [On-line] http://www.tele-graph.co.uk/news/main.jhtml?xml=/news/2007/07/11/ntablet111.xml
3. Ancient Road Publications/Kyle Pope; 2001. *The Seal of Baruch, Jeremiah's Scribe.* [On-line] http://ancientroadpublications.com/Studies/BiblicalStudies/SealofBaruch.html [Accessed 16 June 2016]
4. Testing Worldview/R. Totten/2005, *Archaeology Confirms the Bible is Historical* [On-line] http://worldview3.50webs.com/history.html [Accessed 21 June 2016]
5. *Pontius Pilate's Name Is Found on 2000-Year-Old Ring* (NY Times; 2018) [On-line] https://www.nytimes.com/2018/11/30/world/middleeast/pontius-pilate-ring.html [Accessed 21 January 2019]
6. Josh McDowell, *Evidence That Demands a Verdict* Vol.2 (Campus Crusade for Christ, SanBernadino, CA; 1975), pp. 309-311
7. Israel Finkelstein, Amihay Mazar, Brian Schmidt, *The Quest for the Historical Israel* (Society of Biblical Literature; 2007), p. 14.
8. Christian Evidence Ministries: *Mari Tablets* [On-line] http://christianevidences.org/archeological-evidence/eye-witness-testimony-supporting-old-testament-accuracy/mari-tablets [Accessed 21 June 2016]
9. All About Archeology: *Moabite Stone* [On-line] http://www.allaboutarchaeology.org/moabite-stone-faq.htm [Accessed 21 June 2016].
10. *Ten Top Biblical Archaeology Discoveries* (Biblical Archeology Society) [On-line] https://www.biblicalarchaeology.org/free-ebooks/ten-top-biblical-archaeology-discoveries/ [Accessed 21 January, 2019]
11. *Biblical Archeological Evidence for the Bible* [On-line] http://www.truthnet.org/index.php/25-reasons-to-believe/398-12-reason-biblical-archeology[Accessed 24 January 2019]

12. Ancient Origins, *Justice, Myths, and Biblical Evidence: The Wealth of Information Held in the Ebla Clay Tablets* (Jan. 8, 2017) [On-line] https://www.ancient-origins.net/artifacts-ancient-writings/justice-myths-and-biblical-evidence-wealth-information-held-ebla-clay-021163

13. Watch Jerusalem: *The Top Archelogy Discoveries in 2018* [On-line] https://watchjerusalem.co.il/507-top-discoveries-in-biblical-archaeology-2018 [Accessed 24 January, 2019]

14. *Time*, (March 5, 1990), p. 43

15. On View: Seals of Isaiah and King Hezekiah Discovered (Biblical Archeology Society) https://www.biblicalarchaeology.org/daily/news/seals-of-isaiah-and-king-hezekiah-discovered-exhibit/ [Accessed 21 January 2019]

16. *Ancient City Gate and Shrine from Hebrew Bible Uncovered* (Life Science; 2016) [On-line] https://www.livescience.com/56300-gate-shrine-excavated-in-israel.html) [Accessed 21 January, 2019]

17. Christian Evidence Ministries: *Amarna Tablets* [On-line] http://christianevidences.org/archeological-evidence/eye-witness-testimony-supporting-old-testament-accuracy/amarna-tablets [Accessed 21 June 2016]

18. Ibid. [On-line] http://christianevidences.org/archeological-evidence/additional-archeological-proof/proof-of-wars/

19. Ibid. [On-line] http://christianevidences.org/.../ tower-of-babel-and-other-structures/ [Accessed 21 June 2016]

20. Millar Burrows, op. cit., pp. 1, 42

21. Donald Wiseman, *Digging for Truth*, (Viewpoint no:31; ISCF)

22. Nelson Glueck, *Rivers in the Desert: History of the Negev*, (Jewish Publication Society of America, Philadelphia, PA; 1969), p. 31

23. Frank Harber, *Reasons for Believing, A Seeker's Guide to Christianity* (New Leaf Press, AZ; 1998), p. 70

Science Agrees with the Bible

1. C. S. Lewis, *Miracles* (Harper Collins, London; 1947), p.166

2. Chuck Missler, *The Creator Beyond Time and Space* (Word for Today; 1995), pp. 88-90

3. Asit K. Biswas, *Notes and Records of the Royal Society of London*, (Vol. 25, No. 1; Jun. 1970), pp. 47

4. op. cit. [On-line] http://christianevidences.org/scientific-evidence/aerodynamics/air-pressure-and-generating-lift/

5. National Geographic/James Cameron, *Deep Sea Challenge – The Mariana Trench* [On-line]. http://www.deepseachallenge.com/the-expedition/mariana-trench/ [Accessed 21 June 2016]

6. op. cit. [On-line] http://www.christianevidences.org/oceanography

7. Grant R. Jeffery, *The Signature of God* (WaterBrook Press, Colorado Springs, CO; 2010), p. 115

8. Evolution Facts/Vance Farrel; 2006. [On-line] http://evolution-facts.org/Evolution-handbook/E-H-12a.htm [Accessed 21 June 2016]

9. Lynn Margulis quoted in "UMass Scientist to Lead Debate on Evolutionary Theory," *Brattleboro Reformer*, February 3, 2006.
10. Raymon Tallis *Aping Mankind: Neuromania, Darwinitis, and the Misrepresentation of Humanity* (Acumen Publishing, Durham, UK: 2011), p. 181

Is Jesus a Historical Figure?

1. *The Works of Josephus – The Antiquities of the Jews* (Hendrickson Publishers, Peabody, MA; 1987), 18.3.3, p. 480
2. Ibid., 20.9.1, p. 538
3. John P. Koster, Jr., *The Atheist Syndrome* (Wolgemuth & Hyatt Pub., Brentwood, TN; 1989), p. 61
4. Tacitus, Annals 15.44, cited by Lee Strobel, *The Case for Christ*, (Zondervan Publishing House, Grand Rapids, MI; 1998), p. 82
5. Pliny, Letters, transl. by William Melmoth and W.M. Hutchinson (Harvard Univ. Press, Cambridge, MA; 1935), vol. II, X: 96, cited in Habermas, *The Historical Jesus*, p. 199
6. Lucian, "The Death of Peregrine," 11-13, in The *Works of Lucian of Samosata*, transl. by H.W. Fowler and F.G. Fowler, 4 vols. (Clarendon, Oxford, England; 1949), vol. 4., cited in Habermas, *The Historical Jesus*, p. 206
7. John P. Koster, Jr., op. cit., p. 61
8. David Bercot, *A Dictionary of Early Christian Beliefs* (Hendrickson Publishers, Peabody, MA; 1998), p. 558
9. Ibid.
10. Thomas Aquinas, quoted by Nancy Gibbs, "The Message of Miracles" *Time* (New York, NY; April 10, 1995), Vol. 145, No. 15, p. 68

The Authenticity of the Bible

1. N. L. Geisler, "New Testament Manuscripts" in *Baker Encyclopedia Of Christian Apologetics* (Baker Books, Grand Rapids, MI; 2002), p. 532
2. Patheos/Mark D. Roberts, *Can We Know What the Original Gospel Manuscripts Really Said?* [On-line] http://www.patheos.com/blogs/markdroberts/series/can-we-know-what-the-original-gospel-manuscripts-really-said/ [Accessed 21 June 2016]
3. D. Wallace, "The Gospel According to Bart" (*Journal of the Evangelical Theological Society* 49; June 2006), p. 330
4. Bruce K. Waltke, "*The Reliability of the Old Testament Text,*" Willem A. VanGemeren, ed. *New International Dictionary of Old Testament Theology & Exegesis*, Volume 1 (Zondervan Publishing House, Grand Rapids, MI; 1997), p. 65
5. Josh McDowell, *The New Evidence that Demands a Verdict* (Thomas Nelson Pubishers; 1999), p. 70
6. Ancient Hebrew Research Center/ Jeff A. Benner, *The Great Isaiah Scroll and the Masoretic Text* [On-line] http://www.ancient-hebrew.org/bible_isaiahscroll.html [Accessed 21 June 2016]

7. Testing Worldview/R. Totten/2005, *Archaeology Confirms the Bible is Historical* [On-line] http://worldview3.50webs.com/history.html [2016]

8. Carsten Thiede, The LA Times, Saturday, December 24, 1994

9. Norman L. Geisler and William E. Nix, *A General Introduction to the Bible* (Moody Press, Chicago, IL; 1968), p. 151; updated in 2013 by author [On-line]:http://www.normgeisler.com/articles/ Bible/Reliability/Norman%20Geisler%20-%20Updating%20the%20 Manuscript%20Evidence%20for%20the%20New%20Testament.pdf

10. Norman L. Geisler & William E. Nix. *From God to Us* (Moody Press, Chicago, IL; 1981), p. 157

11. Paul Lee Tan, *Encyclopedia of 7700 Illustrations: A Treasury of Illustrations, Anecdotes, Facts and Quotations for Pastors, Teachers and Christian Workers* (Bible Communications, Garland TX), 1996, c1979, #391

The Uniformity of the Bible

1. C. I. Scofield, *Scofield Study Bible* (Oxford University Press, NY; 1967), p. 328, note 1

Prophecy, the Proof of Inspiration

1. The Berean Call/David Hunt/August 1, 2000, *Who is Jesus?* [On-line] http://www.thebereancall.org/node/5702 [Accessed 21 June 2016]

2. Tentmakers/Atheism Quotes [On-line] http://www.tentmaker.org/ Quotes/atheismquotes.htm [Accessed 21 June 2016]

Worldviews, Religion, and Revelation

1. Sharon Morad Leeds, summary of *The Origins of the Quran: Classic Essays on Islam's Holy Book*, Ibn Warraq, ed. (Prometheus Books, 1998) *http://www.angelfire.com/or/don9840/Quran.html*

2. *Documentary History of the Church* 4:461, June 11, 1843 quoted in Daniel Ludlow, ed. *Latter-day Prophets Speak* (Bookcraft, Logan, UT; 1988)

3. Hal Hougey, *Archaeology and the Book of Mormon*, (Pacific Publishing Co., Concord, CA; 1983), p. 12

4. Joel Kramer and Jeremy Reyes, *DNA vs. The Book of Mormon* (DVD by Living Hope Ministries; 2003)

5. Apologetic Express/Dewayne Bryant, *The Book of Mormon and the Ancient Evidence* [On-line] http://www.apologeticspress.org/ apcontent.aspx?category=11&article=975 [Accessed 21 June 2016]

6. Walter Martin, *Kingdom of Cults* (Bethany House, MN; 1970), p. 64

7. Freeminds. *Translators of the New World Translation* [On-line]:http://www.freeminds.org/history/NWTauthors.htm and https://www.blueletterbible.org/study/cults/exposejw/expose33.pdf [both Accessed 21 June 2016].

8. Towerwatch. *The Failed Prophecies of The Watchtower Bible and Tract Society* [On-line] http://www.towerwatch.com /Witnesses/Prophecies/ failed_prophecies.htm [Accessed 21 June 2016].